AS in a Week

Elizabeth Lilah and
John Milner,
Abbey College, Manchester
Series Editor: Kevin Byrne

you need

SUCCESS OR YOUR MONEY BACK

Letts' market leading series AS in a Week gives you everything you need for exam success. We're so confident that they're the best revision books you can buy that if you don't make the grade we will give you your money back!

HERE'S HOW IT WORKS

Register the Letts AS in a Week guide you buy by writing to us within 28 days of purchase with the following information:

- Name
- Address
- Postcode
- Subject of AS in a Week book bought

Please include your till receipt

To make a **claim**, compare your results to the grades below. If any of your grades qualify for a refund, make a claim by writing to us within 28 days of getting your results, enclosing a copy of your original exam slip. If you do not register, you won't be able to make a claim after you receive your results.

CLAIM IF...

You are an AS (Advanced Subsidiary) student and do not get grade E or above.

You are a Scottish Higher level student and do not get a grade C or above.

This offer is not open to Scottish students taking SCE Higher Grade, or Intermediate qualifications.

Registration and claim address:

Letts Success or Your Money Back Offer, Letts Educational, Aldine Place, London W12 8AW

TERMS AND CONDITIONS

1. Applies to the Letts AS in a Week series only
2. Registration of purchases must be received by Letts Educational within 28 days of the purchase date
3. Registration must be accompanied by a valid till receipt
4. All money back claims must be received by Letts Educational within 28 days of receiving exam results
5. All claims must be accompanied by a letter stating the claim and a copy of the relevant exam results slip
6. Claims will be invalid if they do not match with the original registered subjects
7. Letts Educational reserves the right to seek confirmation of the level of entry of the claimant
8. Responsibility cannot be accepted for lost, delayed or damaged applications, or applications received outside of the stated registration/claim timescales
9. Proof of posting will not be accepted as proof of delivery
10. Offer only available to AS students studying within the UK
11. SUCCESS OR YOUR MONEY BACK is promoted by Letts Educational, Aldine Place, London W12 8AW
12. Registration indicates a complete acceptance of these rules
13. Illegible entries will be disqualified
14. In all matters, the decision of Letts Educational will be final and no correspondence will be entered into

Letts Educational
Aldine Place
London W12 8AW
Tel: 0208 740 2266
Fax: 0208 743 8451
e-mail: mail@lettsed.co.uk
website: www.letts-education.com

Every effort has been made to trace copyright holders and obtain their permission for the use of copyright material. The authors and publishers will gladly receive information enabling them to rectify any error or omission in subsequent editions.

First published 2000

Text © John Milner & Elizabeth Elam 2000
Design and illustration © Letts Educational Ltd 2000

British Library Cataloguing in Publication Data
A CIP record for this book is available from the British Library.

ISBN 1 84085 368 9

Prepared by *specialist* publishing services, Milton Keynes

Printed in Italy

Letts Educational Limited is a division of Granada Learning Limited, part of the Granada Media Group

15 minutes

Test your knowledge

1 How do the changing seasons affect the amount of water lost by evapotranspiration?

2 During a rainstorm, what factors encourage overland flow rather than infiltration?

3 What determines whether a river erodes vertically or laterally?

4 Define 'river discharge'.

Answers

1 In winter, loss of leaves reduces transpiration and falling temperatures reduce evaporation rates. Evapotranspiration is highest in summer.
2 Overland flow is more likely to happen when: infiltration capacity has been lowered by urbanisation, deforestation, soil compaction or the existence of a hard pan in the soil; there is a high water table due to antecedent rainfall; ground is frozen in winter; there are steep slopes; rainfall is very intense.
3 In the upland, or mountain stage, streams have high energy due to steep slopes and so vertical erosion creates potholes and gorges. In the lower course of the river, energy is lower due to shallow gradients and so the river erodes laterally, (side to side) and the channel widens.
4 This is the amount of water in the channel in a certain period of time, usually cubic metres per second (cumecs). Remember that velocity multiplied by cross-sectional area gives discharge. A hydrograph shows the way a channel's discharge changes over time.

✔ **If you got them all right, skip to page 10**

River Basins and Hydrology

45 minutes

Improve your knowledge

1

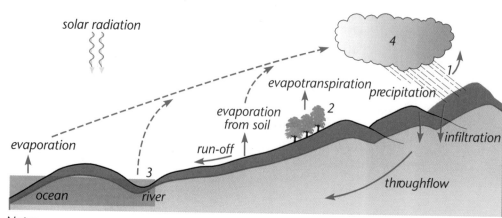

Notes:

1 evaporation can occur during precipitation
2 secondary interception can occur as water drips from trees to shrubs
3 precipitation may fall directly into the river (channel precipitation)
4 evaporated water vapour is condensed into water droplets in the cloud

During interception, precipitation can hit vegetation before the ground and can flow down the stems of plants (stemflow); drip from the surface of leaves (throughfall) or evaporate back (interception loss).

Evapotranspiration is a combination of evaporation of water and transpiration through the stomata of the leaves. Rate of loss is dependent on the season (many species lose leaves in winter), and on the type and density of vegetation. Loss by evaporation alone is increased by higher temperatures, high wind speeds and low humidity.

2 Infiltration occurs when water enters the ground. As the rain continues, infiltration capacity may be reached and water stays as surface storage or, if on a slope, overland flow (run-off). Deeper percolation results in throughflow within the soil layers. Water may eventually enter permeable rock to become groundwater.

Streams begin as tiny rills on a slope, which join to form larger gullies. These join small tributaries and thus enter the main channel.

The amount of rain intercepted in a forest may be calculated by placing rain gauges underneath the trees and in open space nearby. The difference between the average amounts collected is the amount intercepted.

River Basins and Hydrology

Wide, shallow channels develop where the banks are easily eroded. If the banks consist of more cohesive material such as silt and clay, then the channel may be deeper and narrower. Channels which have cut through rock are usually deep and narrow, and are called gorges. Many channels, especially in the lower course, are sinuous and meandering, with numerous river bends from side to side. Some stream channels, especially glacial outflow streams, have braided channels, where the stream splits into numerous different channels separated by sediment bars.

River load comes from the eroded bed and banks of the river and its tributaries. The river's capacity to transport load depends on how much energy it has, given by its velocity of flow. This energy can either erode or transport material. If the river's energy (velocity) falls, deposition occurs. This is shown in the diagram below:

River channel depth can be measured with a ruler held vertically. Readings are usually taken every 50 cm across the width of the river. River gradient is measured using two marker poles and a clinometer, which measures angle of tilt.

Stream competence is the power of a stream as shown by the largest sized particle being moved as bedload.

The Hjulström Curves

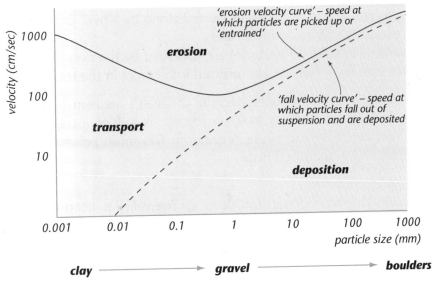

Note: Tiny clay particles are more difficult to pick up (entrain) than larger gravel particles as they are highly cohesive.

The velocity needed to transport particles in suspension is less than that required to entrain them.

There are four types of river load:

- bedload – large particles rolled along the bed – even boulders at times of high discharge

River Basins and Hydrology

- *saltation load – medium particles bounced along the bed*
- *suspended load – small particles permanently suspended in the water*
- *solution load – minerals dissolved in the water.*

3 Channel processes

Upland river stretches often have turbulent flow with numerous eddy currents caused by high channel roughness. Further downstream, flow is less turbulent. Sometimes laminar flow occurs in a very smooth channel, with layers of water flowing over one another.

River velocity is increased by steep slopes (such as in the mountain stage), clear channels (no boulders or debris) and channels with a high efficiency (measured by hydraulic radius).

Erosion takes four forms:

- attrition – load rubs together and becomes smaller
- abrasion – the grinding action of materials carried along by a river creates potholes and wears away banks and river beds
- corrosion – minerals, (especially limestone) are dissolved by the river
- hydraulic action – bubbles of air are compressed into cracks in the rock

Deposition occurs when energy is reduced because the river's gradient decreases in its lower stages or it enters a lake or the sea. A sudden increase in load, perhaps due to a landslide, can cause deposition. The sorting of materials occurs as larger particles are deposited first.

Erosional features

In the mountain stage, the river appears fast-flowing. The valley is steep-sided and V-shaped. The valley floor has a steep, uneven gradient with potholes, waterfalls and rapids. Vertical abrasion is the main process, creating potholes which join together to lower the bed. The river flows from side to side around interlocking spurs of land. Waterfalls or rapids occur where there are hard bands of rock in the valley.

As the gradient becomes less steep, the interlocking spurs have been eroded back to form bluffs, and lateral (sideways) erosion widens the valley floor and allows the river to meander. Erosion and deposition occur simultaneously due to the differences in velocity in the meander bend.

Solution load can be measured with a conductivity meter. Two electrodes are held in the stream. The greater the conductivity of the water, the greater the concentration of dissolved minerals.

HR = Cross-Sectional Area / Wetted Perimeter

River Basins and Hydrology

Rejuvenation of a river happens when the sea level falls or occasionally if the land surface rises due to tectonic activity. The river's energy is increased and it begins to erode vertically downwards once more, forming terraces as it cuts into the floodplain or incised and ingrown meanders in areas of resistant rock.

Depositional features

When the river is in its floodplain stage, the meanders do not fill the valley floor and are irregular. Oxbow lakes may form, which are stagnant and cut off from the main channel. Each time the river floods, natural embankments called levées are built up. If the river flows into a lake or the sea, a delta may form as the sudden drop in energy causes deposition. Deltas require large quantities of load and minimal wave erosion.

4 Discharge ('Q')

This is the amount of water in the channel in a certain period of time, usually cubic metres per second (cumecs). Remember that velocity multiplied by cross-sectional area gives discharge. A hydrograph shows the way a channel's discharge changes in response to a period of precipitation.

$Q = V \times A$

A typical storm hydrograph

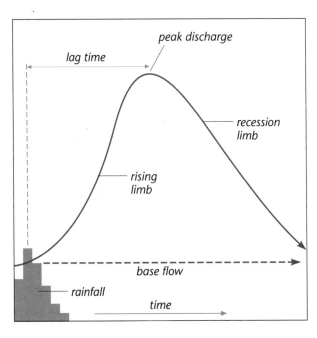

Notes:
1 steepness of rising limb depends on type of rainfall and basin characteristics
2 steepness of recession limb depends on permeability of soil and the type of vegetation or land use
3 lag time = number of hours/days between peak rainfall and peak discharge; low infiltration, rapid run-off and high drainage density reduce lag time and steepen limbs

Discharge can be measured using a 'V-notch weir', a wooden board with a 90° notch which dams the stream and lets the water flow at a measurable rate.

A long, narrow basin with gentle slopes, dense woodland and few tributaries would have a hydrograph with shallow limbs and a long lag time. A rounded basin with steep slopes, saturated soil, sparse shrub cover and high drainage density would have a hydrograph with steep limbs and a short lag time.

In urban areas, impermeable surfaces produce fast run-off and very low infiltration. There is very low evapotranspiration and low interception due to lack of vegetation and so water reaches stream channels quickly via smooth, efficient drainage systems.

Urban channels are artificially smoothed.

If a basin (catchment area) is deforested, reduced interception drastically reduces infiltration capacity and evapotranspiration is greatly lowered. Greatly increased run-off and possible flooding will result.

River regimes are sometimes called 'flow hydrographs' and represent the variation in a river's discharge over a whole year. The main influence on this is the annual distribution of precipitation:

- Climates with a winter maximum of precipitation (such as the UK) will show highest discharge in the winter. Low temperatures also mean that loss from evapotranspiration is lower at that time of year.
- Rivers in equatorial regions such as the Amazon have little variation in discharge over the year, as rainfall remains high and regular all year.
- Many river regimes show a peak discharge in spring when water from snowmelt enters the channel, called ablation. This can affect many rivers as even those in areas of low snowfall may receive tributary water from streams in colder, upland areas.
- Rivers in very cold polar and tundra regions freeze up all winter, when their discharge is zero. They have a very high peak in the spring and early summer and then discharge falls to zero again in early winter.

The discharge pattern shown in a flow hydrograph reflects the annual pattern of precipitation but also that of temperature.

Flooding occurs when the river is over bankfull. Floodplains exist naturally in the lower stages of a river – flooding becomes a hazard when man intervenes by settling on the floodplain.

Causes are numerous and each flood has a combination of them. Heavy, prolonged rainfall with antecedent rainfall already having saturated the soil is the primary cause. Sun-baked ground or urban surfaces, deforestation, silting up of channels due to soil erosion, sudden snowmelt in spring, dam failures and earthquakes can all be causal factors. The dumping of rubbish in channels causes many small urban floods.

River Basins and Hydrology

Some typical river regimes

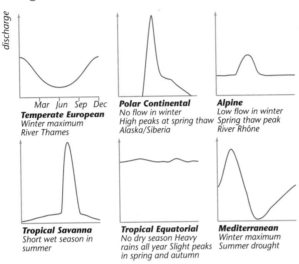

Temperate European
Winter maximum
River Thames

Polar Continental
No flow in winter
High peaks at spring thaw
Alaska/Siberia

Alpine
Low flow in winter
Spring thaw peak
River Rhône

Tropical Savanna
Short wet season in summer

Tropical Equatorial
No dry season Heavy rains all year Slight peaks in spring and autumn

Mediterranean
Winter maximum
Summer drought

Experts use flood recurrence intervals to plan the management of large, complex river basins such as the Nile, Mississippi or Thames. Flood protection measures include modifications to banks, building higher levées, removal of boulders and reducing the sinuosity of the course. Regular dredging, construction of 'floodway' channels or reducing variability of discharge with a dam all reduce the likelihood of flooding by increasing the velocity and channel efficiency, thus allowing the water to flow away quickly and safely.

Flood abatement tackles the problem at source by increasing the lag time, allowing the river system to cope with a slower, more gradual input of water. Afforestation in the basin increases interception and evapotranspiration so reducing and slowing run-off. Terracing of hillsides and ploughing along contours slows and reduces run-off by increasing infiltration.

River basin schemes often adversely affect local communities: some types of employment will grow, some will decline. How will the local natural environment change? The benefits of improved agriculture will not reach everyone and new sources of hydroelectric power may attract large industries. In each case there will be different groups and individuals for and against.

The use of computer modelling of floods, improved precipitation forecasts, better warning systems, modifications to buildings and infrastructure and improved evacuation procedures are important in reducing the effects of flooding.

You should have studied the flood protection measures and channel modifications for a particular river as a case study.

List the reasons why floods cause more deaths in LEDCs.

River Basins and Hydrology

Use your knowledge

Hint

1 What effects does vegetation have in the hydrological cycle?

Consider how vegetation protects the soil.

Always carefully study graphs and diagrams for a few minutes before answering.

2 The hydrographs in the diagram below show how two different rivers respond to the same rainstorm:

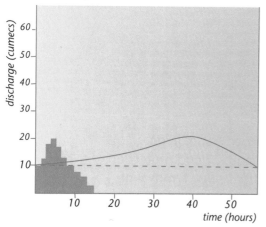

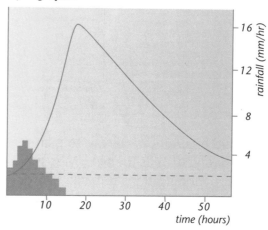

a) What is the lag time for river B?
b) What is the approximate baseflow for river A?
c) Compare and contrast the two hydrographs.
d) What factors may cause the long lag time in hydrograph A?
e) If the drainage basin for hydrograph A was rapidly urbanised, describe and explain how the lag time and peak discharge would be affected.

In question c, you must make comparative statements, not just separate descriptive statements for each graph.

Divide the question into the two obvious halves.

3 Discuss the importance of people in affecting the degree and frequency of flooding using a major river basin that you have studied.

15 minutes

Test your knowledge

1 Look at a map of the Pacific Ocean. Why is Hawaii good for surfing?

2 Why do headlands erode relatively quickly, considering that they are comprised of hard rock?

3 Where and why do spits from?

4 Why is the use of groynes controversial?

Answers

1 Very isolated – very long 'fetch' – large waves. 2 Because of wave refraction, where the shallower water around a headland 'refracts' or turns the waves into it. In addition, whereas a straight coast has an 'angle of attack' of only 180°, a headland is open to attack on all sides, having an angle of attack of over 300°. 3 Beach material is carried along the coast by the process of longshore drift. Deposition of this sand and shingle occurs due to lower energy, where the coastline changes direction or where a river enters the sea. This forms a long bank of material jutting out to sea from the coast. The direction of the prevailing waves governs the shape of the spit. 4 They stop/reduce longshore drift and so stop accumulation of material further down the coast where it might be needed to maintain beaches for recreation or protection.

 If you got them all right, skip to page 19

Coastal Environments

Improve your knowledge

Coastlines vary considerably in their type and character. They range from the steeply shelving coasts of mountainous areas to the gentle profiles of lowland coasts.

Where the coastal rock is resistant, cliffs and bays occur. On softer rock, flat saltmarshes and mudflats are common.

 Marine energy

Energy transports material to and along the coast and attacks and erodes coastal rocks. This energy comes from the development of waves. Waves are increased in size and energy by increases in wind velocity, wind duration and length of fetch (stretch of open water).

Dominant waves (constructive or destructive) are the type that has most effect on a certain coastline. Prevalent waves are those which affect the coast for most of the time.

Waves 'break' when the water depth is equal to their wave height. Swash flows up the beach and backwash down.

A 'breaker' can be either:

- plunging and destructive: high frequency, short swash, long backwash, more material eroded
- spilling and constructive: low frequency, long swash, short backwash, more material deposited.

Tidal range is the difference between high and low water mark. High tide is in spring when the sun, earth and moon are aligned. Neap (lowest) tide is when there is no alignment. Tides can be classified as macro (range over 4m), meso (range 2-4m) or micro (less than 2m).

Storm surges are very fast rises in sea level. Seawater is pushed up again a stretch of coastline causing severe inland flooding. The consequences are even more disastrous when the coast is heavily populated. The southern part of the North Sea is particularly prone to storm surges when atmospheric depressions cause the sea level to rise, which may then be turned into high storm waves by the strong winds that are associated with such depressions.

Many fieldwork techniques are straightforward to describe. The dimensions of waves (height and length) are just observed and then estimated using a long ruler. It is important to take an average from many readings.

2 Coastal processes

Coastal systems are the result of the interaction of several processes including geological, weathering, erosional, depositional and biotic processes. Along most coastlines, human activity modifies and disrupts these processes.

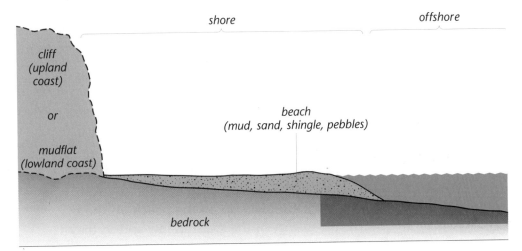

Cliffs can be surveyed to find out how the cliff type changes along a coastline.

Processes of erosion acting on the cliff face:

Top of cliff:
- mechanical, chemical and biological weathering
- salt splash erosion

Mid cliff:
- hydraulic action (wave pounding)
- pneumatic action (air forced into cracks under pressure)
- corrosion (waves dissolve limestone cliffs)

Lower cliff:
- corrasion (waves 'armed' with pebbles)
- attrition (cliff fragments are rounded into pebbles)

Deposition occurs when accumulation of material (sand, pebbles) is greater than its depletion, usually in sheltered areas with low energy waves.

Consider how the presence of salt water increases some types of weathering.

Features of coastal erosion:

- **Cliffs**: shape dependent on rock type and structure. Steep limestone and granite cliffs due to slow erosion; many bays and headlands due to faster erosion where rock is less resistant (such as shale). Waves refract either side of headlands because water is shallower.
- **Arches, stacks and stumps**: refraction around resistant headland erodes a cave through to become an arch. Roof collapse creates a stack. Further erosion (pneumatic and hydraulic action and corrasion) leaves a stump.
- **Caves and blowholes**: Pneumatic action enlarges crack into a cave; erosion continues to top of cliff.
- **Wave-cut platforms**: Cliff is undercut by lower-cliff processes to form wave-cut notch. Cliff above collapses leaving platform.

Learn which specific erosional processes act upon the coast to form each feature.

3 Features of coastal deposition

Beaches

Constructive waves cause sand and shingle to build up on the beach. As these flat, low waves break on the beach, the energy in the swash is low and so they have a weak backwash. The beach gradient or steepness is increased and berms occur – small ridges of shingle marking high tides. Destructive waves are steep and so high in energy. When they break, the strong backwash carries material down the beach, reducing its gradient.

On shingle beaches, coarse material allows water to percolate through and so backwash is limited and shingle builds up on the beach. At the top of the beach a storm beach may form, made up of larger pebbles and boulders rolled up by the larger waves. On sandy beaches, the material does not allow water to pass through as easily and so there is considerable erosion by the backwash. Sandy beaches are likely to be less steep.

If waves break obliquely to the shore, the backwash is still at right angles to it and so material is transported along the beach (longshore drift).

The relationship between beach particle size and beach gradient could be tested for fieldwork. Average particle size could be determined by random sampling then measured against gradient using Spearman's Rank Correlation Coefficient to see if a correlation exists.

Longshore drift reduced by the use of groynes

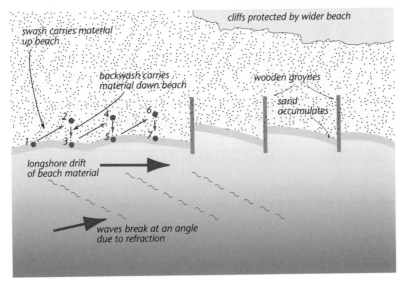

Spits form where a river enters the sea or where the coast changes direction (loss of energy). Material carried down the coastline by longshore drift builds up and is shaped by the direction of dominant waves. Spits may have small, lateral spits behind the main spit in a salt-marsh area and a hooked distal end.

If the spit grows to reach an island a tombolo is formed. It may grow across a bay to from a bay bar, enclosing a lagoon. If a large, triangular spit forms in deeper water it is called a cuspate foreland.

Mudflats and salt marshes occur in sheltered water if deposition exceeds river erosion in river estuaries. They have many small creeks at low tide and are quickly colonised by halophytic (salt-tolerant) plants.

Sea level changes
Changes in base sea level can be eustatic (worldwide change due to change in actual sea level) or isostatic (local change due to land-level change).

During ice ages, more water was stored at the poles and so there would have been an eustatic fall in sea level. At the same time, isostatic rises in sea level would have occurred due to the land 'sinking' under the weight of the ice. When the ice retreated, the land could rise due to the release of pressure causing isostatic falls in sea level.

You should be able to draw simple labelled sketches of each feature.

Make sure you are clear about the way that uplifting land can cause falling sea levels and vice versa.

Coastal Environments

Tectonic processes can cause the land to uplift (orogeny) and hence cause falling sea levels. Localised land tilting (epeirogeny) can cause coastlines to dip and be inundated by the sea.

Raised sea level creates submerged upland coasts with rias and fjords or submerged lowland coasts with broad, shallow estuaries and mudflats.

Fjords are formed by rising sea levels, often at the end of the last ice age. They are glaciated U-shaped valleys in upland or mountainous areas, which run perpendicular to the coast. When sea levels rose, these coastlines became submerged as the valleys were inundated. Some of the best examples are found along Norway's west coast.

Lowered sea level creates emerged upland coasts with a steeply rising shore and raised beaches or emerged lowland coasts with coastal plains.

In the future, sea levels are expected to rise due to the continuing impact of global warming. This will occur because of:

- melting mountain glaciers releasing stored water;
- oceanic thermal expansion due to higher temperatures;
- some melting of polar ice caps – although this will be offset by higher polar snowfall.

Although sea levels may only rise by up to one metre, huge areas of low-lying coastal land may be inundated. Such areas are often agriculturally productive and highly populated, such as coastal Bangladesh. Many major world cities, including London, Calcutta and Tokyo, would also be affected.

4 Coastal management

The human impact on coasts is one which creates significant debate and controversy due to the effects further down the coast of interference, for example by the disruption of longshore drift with use of groynes.

Many coastal settlements are threatened by cliff retreat, and sea walls, revetments and breakwaters are used to slow down the process. Coasts are also used extensively for recreation and the management of this must be considered.

Many coastlines are left unprotected as they have little commercial value. Coastlines that have extensive residential settlement, have valuable agricultural

The best way to revise features such as these is to draw a simple labelled diagram for each.

You may be asked to write clear explanations of the causes of the greenhouse effect and the resulting global warming. What measures have been taken by governments? How successful have these been?

land or are used extensively for recreation are protected by coastal defence measures. These fall into two categories:

- 'hard' solutions involving engineering schemes;
- 'soft' solutions which supplement natural beach protection.

Governments and local authorities have to weigh up the cost of sea defences against the 'value' of the coastline in a particular area. How this 'value' is worked out is controversial.

	Solution	Comments
Hard solutions	Concrete sea wall deflects wave energy	Wall itself can be undercut leading to collapse. Deflected waves erode beach instead.
	Pre-cast concrete panels are bolted onto the cliff face.	Regarded as being ugly, they are unpopular, especially in tourist areas.
	Stone boulders/revetments absorb wave energy	Regarded as being ugly, they are unpopular, especially in tourist areas.
	Improved cliff drainage reduces slumping	Holes are bored into the cliff and water pumped out at the base.
Soft solutions	Groynes retain beach sediment, widen the beach and thus reduce wave energy	Again considered ugly, groynes reduce the amount of material received down the coast, causing beach depletion there.
	Offshore breaks cause the waves to break earlier, thus reducing their impact	Constant maintenance and replacement is expensive.
	Beach nourishment	Adding coarser sand protects the beach and reduces its gradient.

Most coastlines have significant pressures from different sources. These may include:

- visiting tourists and day-trippers – coasts that are within easy reach of large population centres receive most visitors. This pressure is highest in the summer and during school holidays. Car parks and toilet facilities are necessary but may spoil the area's natural beauty. Footpaths to and from the beach become eroded.

- agriculture – overstocking with cattle and sheep can cause dune areas to be stripped of vegetation and the sand can be blown away. Agricultural grasses can take over dune areas, displacing the natural flora.
- sand and shingle extraction – many beaches have beach material extracted by mechanical diggers for use in the construction and road building industries. This lowers the beach level, which brings the sea closer and removes the source of sand which is necessary to maintain the sand dunes.

Pressures and conflicts of interest may occur within each group. For example, windsurfers may come into conflict with anglers or bird watchers.

You will need detailed case study knowledge, including a sketch map, of the issues and possible solutions for a particular stretch of coastline.

Coastal Environments

Use your knowledge

1 Discuss the interrelationships that exist between deposition and erosion along coastlines.

2 Describe the other factors that affect the form of sea cliffs along with wave action.

3

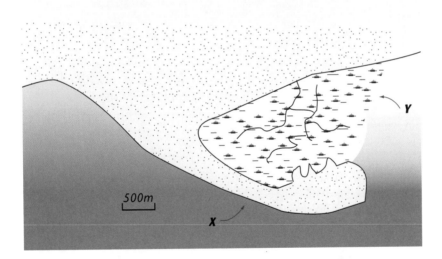

500m

Y

X

a) Name coastal feature X and area Y.
b) Describe area Y.
c) Describe the vegetation that would be typical of area Y.
d) Explain how area Y would slowly change over time.

4 Why do changes in the relative level of land and sea occur? With reference to examples, discuss the human response to the coastal landforms that result from such changes.

Hint

Interactions between the two must be described.

Try to refer to some named cliffs that you have either visited or that you have looked at as a case study.

You need to discuss the area's ecosystem, and its flora and fauna in detail.

There are no marks for explaining the formation of the landforms so don't waste your efforts doing this.

15 minutes

Test your knowledge

1 How does the first trophic level produce biomass?

2 What are the three nutrient stores and what factors influence their relative sizes and the rate of flow between them?

3 The first plants to colonise a newly exposed surface form the _____ _____.

4 What effect does an arresting factor have on a prisere?

Answers

1 Primary producers are the plants, ranging from microscopic algae to large trees. They grow and so increase biomass by using sunlight and carbon dioxide to produce growth by photosynthesis. **2** The three nutrient stores are biomass, litter layer and soil. Their size depends on the area's temperature and precipitation. In the tropical rainforest, the soil store is low due to high leaching because of the high rainfall; litter store is low due to rapid decomposition due to high temperatures. Most of the nutrients are stored in the biomass itself. **3** Pioneer community **4** It stops the prisere from reaching CCC, so that it reaches a plagio-climax instead; this has fewer species and lower biomass.

✔ **If you got them all right, skip to page 24**

Ecosystems

Improve your knowledge

When studying ecosystems you must be aware of the interdependence of life forms, the stability or instability of the system (effect of man) and the relationship of plants and animals to the system.

1 An ecosystem is an interacting community of flora and fauna together with the abiotic (not living) environment which includes the soil, climate and relief. Ecosystems vary in size from a puddle to a large climatic region.

Food chains develop, in which each stage or trophic level is occupied by a different group of organisms. Primary producers (plants) use the sun's energy to photosynthesise and produce biomass. This is consumed by herbivores (primary consumers) which are in turn consumed by carnivores (secondary and tertiary consumers). At each level, energy is lost and so each trophic level has fewer individuals in the population.

2 Nutrients used by plants are stored in the biomass itself, in the soil or in the litter. The size of each store and the rate of flow of transfers between stores is dependent on temperature and precipitation. In the boreal forest, the deep litter layer is the largest nutrient store as low temperatures discourage humus incorporation into the soil by worms (low soil store) and biomass store is low as there is little undergrowth.

Vegetation can be described in terms of its species composition and its layer structure. A plant community may be species-rich (tropical moist forest) or species-poor (tundra).

3 A prisere is the series of stages of colonisation by vegetation on an untenanted site such as a fresh sand dune. The first plants to establish themselves form the pioneer community. With each stage, or sere, species diversity, biomass and soil fertility increase until the climatic climax community (CCC) exists, in equilibrium with the abiotic environment.

4 When an arresting factor (e.g. fire, deforestation, grazing) prevents the CCC from being reached, a plagio-climax exists with fewer species and lower biomass. If the arresting factor is removed, a secondary succession may result in the CCC being reached.

Most questions will ask you to discuss the human impact on the ecosystem. Some will also ask about solutions to the problems.

Easy marks for learning definitions!

These nutrient stores are often shown as three circles of different sizes. The relative values of the flows between them are shown by arrows of variable thickness.

Very few areas have their true climatic climax vegetation due to interference by man.

Hydroseres

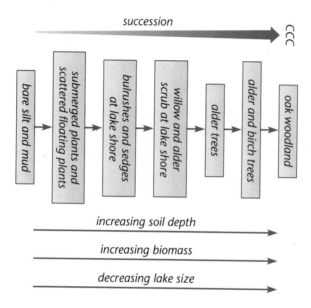

Lakes are often poor in nutrients and so early colonisers will be only mosses and algae. These will attract insects and the biomass of the area begins to increase. Water-tolerant plants will grow under water or floating on the surface and marsh grasses trap and build up sediment until a thin soil exists which can support shrubs and small trees.

The hydrosere is often affected by human actions, especially in areas such as the Lake District or the Norfolk Broads, which receive high numbers of tourists. Lake shores may be damaged by passing speedboats, fishing wire can harm fauna and visitors may drop litter and disrupt delicate shore-edge ecosystems.

The human effect on different priseres and ecosystems must be backed up with case study knowledge.

Haloseres

Similar to a hydrosere, a halosere exists in the area where water meets land. In river estuaries, silt is deposited by the incoming tide on wide mudflats and saltmarshes. The first colonisers can tolerate being submerged in salt water for over twelve hours at each high tide. The material deposited by the tide as it retreats is increased by river load brought into the estuary by in-flowing streams.

Ecosystems

First Stage	Green algae grows which can tolerate submergence. Algae traps mud and sand which slowly accumulate.
Second Stage	*Halophytes (salt-lovers) such as spartina colonise the water's edge. They are above water some of the time. These trap more sediment.*
Third Stage	Grasses start to form a more continuous cover. They are only submerged for some of the time.
Fourth Stage	As the soil depth increases, non-halophytes which are never submerged begin to colonise. These include small *ash* and *alder* trees.

Estuaries are often favoured sites for industry due to the cheap flat land which is close to the sea for transport. They often receive much industrial pollution and sewage from urban areas. This can cause eutrophication, when algal growth and oxygen depletion can kill fish and plant life. Heavy metals such as cadmium and mercury can be consumed by molluscs and passed up the food chain by the process of bioaccumulation.

The study of a hydrosere, halosere or other prisere would require some form of vegetation sampling. Sampling should be random, but this may leave large areas unsampled. A systematic sampling method can avoid this, where the area is first divided into squares.

Psammoseres

A psammosere develops on a sand dune:

First Stage	*Lyme grass and marram grass colonise the dune. These have to survive arid conditions, as the dune cannot hold rainwater. Some species can fold their leaves to reduce transpiration loss. Some have very long tap roots.*
Second Stage	*Rotting plants add humus and nutrients. Other plants such as heather appear. Their roots bind and hold the sandy soil.*
Third Stage	*Reeds and taller shrubs begin to grow.*
Fourth Stage	*Small trees such as ash and hawthorn begin to grow.*

It is important that you learn two or three plant or shrub names associated with each prisere that you have studied.

Sand dune areas are often affected by visiting holidaymakers and tourists who may damage the area by leaving litter, building fires and eroding footpaths. These problems are worse in the busy summer months, when areas may have to be fenced off. Facilities such as car parks, toilets and information points may be provided to encourage visitors to all visit the same area and so limit any damage to that one area.

Ecosystems

45 minutes

Use your knowledge

1 Draw a simple diagram to show the stores and flows of nutrients in the tropical rainforest. Explain the relative sizes of the stores and flows.

2 Describe three ways in which deforestation could disrupt this nutrient system.

3 Study the two diagrams.

Northern coniferous forest (taiga)

L = litter
B = biomass
S = soil

Temperate grassland (prairie/steppe)

a) Define the terms 'litter' and 'biomass'.
b) Explain the processes by which leaf litter becomes incorporated into the soil.
c) Why does the size of the largest store vary in each?

Hint

Consider the temperature and precipitation patterns.

Consider how deforestation would alter the hydrology of the area. How would this affect nutrient flows?

Some questions do not use thicker arrows to show larger flows so you cannot comment on the relative sizes of flows.

Easy marks for definitions!

Consider differences in climate.

15 minutes

Test your knowledge

1 State the differences between continental and oceanic crust.

2 How do the secondary effects of earthquakes contribute to a high death toll?

3 How does rock resistance affect the type and rate of weathering?

4 State the differences between slides and flows.

Answers

1 Continental crust is thicker (up to 75km) and it is older and more dense and made of mostly basalt. Oceanic crust is only 6–11 km thick, comprised of mainly silica and basalt. 2 Fires, floods, water-related diseases and famine may all occur after the quake. Fires are often fuelled by broken gas pipes and cannot be extinguished if water mains have burst. 3 Hard, fine-grained rock with few minor structures weathers slowly as it cannot be penetrated by water. Light-coloured rock reflects the sun's heat and so is more resistant. 4 Slides are more rapid and have less internal deformation. The material moves downslope in a single block rather than breaking up as in a flow.

 If you got them all right, skip to page 32

Tectonics and Surface Processes

45 minutes

Improve your knowledge

1 Plate tectonic theory **developed from Wegener's theory of** continental drift. **The evidence for this is:**

- the 'jigsaw fit' of the continents;
- the fossils of mesosaurus which were found only in southern Africa and Brazil;
- the geological match between rocks in North America and Europe.

Have a look at this 'jigsaw fit' on a world map in an atlas or textbook.

In 1948, the Mid-Atlantic Ridge **was discovered, a continuous mountain range linking Iceland in the north with Tristan da Cunha Island in the south.**

It was found that rock in this ridge is young compared to that at the continental margins and that the Atlantic is widening by approximately 5 cm a year due to this 'sea floor spreading'.

The parallel bands of rock either side of the ridge revealed that iron mineral particles aligned themselves to the pole as the lava cooled. As these rocks were dated, it was discovered that the regular reversals of the earth's magnetic field **were shown in these alternate rock bands and that these are almost symmetrical either side of the ridge (**palaeomagnetism**).**

The Mid-Atlantic Ridge is an example of a constructive **or** divergent **plate boundary, where two oceanic plates move apart. Magma constantly rises up to fill the opening, building up into an** oceanic ridge. **The peaks of such ridges are** submarine volcanoes **and may grow above sea level (e.g. Surtsey near Iceland).**

Study a map of the plate boundaries in your textbook. Learn two or three of its main features.

When there is lateral movement, transform faults **are produced** at right angles to the plate boundary.

Tectonics and Surface Processes

A Constructive Plate Boundary

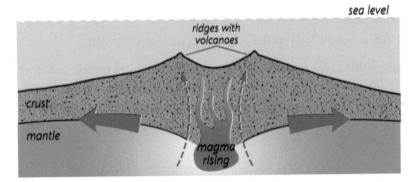

At destructive boundaries, denser oceanic crust is forced under continental crust, (e.g. along the Pacific coast of South America). Here, a long oceanic trench is formed in the subduction zone and the subducted oceanic crust melts due to the heat from intense friction.

This magma forces its way to the surface to form volcanoes and fold mountains on land (the Andes) or island arcs at sea (Japan).

At conservative plate margins, two plates slide past each other and crust is neither created nor destroyed. The two plates move horizontally alongside each other along a huge transform fault (San Andreas in California). Violent earthquakes result from the sudden jolt or slippage.

Draw simple labelled diagrams for the destructive and conservative margins.

Volcanoes are often classified by their shape:

- Shield volcanoes are made of basic, runny lava and so have gently sloping sides.
- More viscous acid lava forms steeper dome shapes.
- Some are comprised of alternate layers of ash and lava or cinders and lava.

2 Earthquakes are caused by a build up of pressure within the crust. Its sudden release creates shock waves or seismic waves, concentrated on a surface epicentre and on a subsurface focus.

The plate margins are often heavily populated because:

- People perceive the risk of earthquake as being minimal compared to the day-to-day problems of their lives.

- Many volcanic areas have very fertile soil or may contain valuable minerals such as gold and copper.
- Dormant volcanoes provide good defensive sites.
- People have the technology and money to cope with the hazard.

As well as the danger from lava, volcanoes also eject poisonous gases, ash and rocks and can cause floods, pyroclastic clouds (clouds of hot gas, molten rock and ash) and huge mudflows (lahars). Often the lava is slow moving. In this case people have plenty of time to evacuate, but cannot stop the lava destroying their homes.

Earthquakes usually kill far more people due to resultant fires, floods and disease (especially water-related diseases) and starvation rather than collapsing buildings. In the very long term, whole communities can disappear as people are afraid to return to the area and harvests are disrupted.

Earthquakes at sea can cause large tsunamis or giant waves, which drown thousands and destroy harvests. They are generated by movements in the ocean floor, either directly due to an earthquake, or due to slumping of undersea sediments near tectonic subduction trenches. The high waves can travel thousands of kilometres before reaching land and flooding coastal areas. Small tsunamis caused by minor earthquakes can be seen quite frequently as unusually large 'freak' waves which travel further up the beach.

In many more economically developed countries (MEDCs, such as Japan, buildings can be reinforced with cross bracings and flexible gas mains can be used to reduce the risk of fire. Some buildings have alarms that cut off the gas supply automatically and windows and furniture may be adapted to reduce injuries. To cope with ground liquefaction, buildings are constructed on floating concrete rafts.

Although the location of earthquakes can be predicted, the timing of them cannot yet be anticipated with much accuracy. Volcanoes are sometimes difficult to predict and may occur explosively with no warning.

3 Weathering is the process of rock denudation, in situ, by natural agents. There are three types:

- mechanical/physical – loosening and cracking;
- chemical – rotting and chemical decomposition;

Consider the different impacts of volcanoes and earthquakes in MEDCs and LEDCs

Write a similar paragraph describing the measures that could be taken in a LEDC

Learning simple definitions gets easy marks. Say them out loud thirty times – it doesn't take as long as you'd think.

- biological – loosening by tree roots and burrowing animals and acidification of soil by chelation of humic acid from rotting organic matter.

Mechanical processes:
- exfoliation – diurnal (daily) heating and cooling fractures surfaces, aided by small quantities of moisture;
- freeze-thaw – water in cracks freezes and expands, cracking the rock. Diurnal freeze-thaw cycle is most effective;
- granular disintegration – individual grains are broken off sedimentary rock;
- pressure release – intrusive igneous rock cools at great pressure but expands and is weakened when exposed to the atmosphere;
- salt crystallisation – salts enter cracks in solution, then dry, crystallise and expand – widening the cracks.

Chemical processes:
- oxidation – minerals change into looser, softer oxides by reacting with oxygen in the air;
- hydration – minerals absorb water and expand so stressing the rock, (anhydrite changes to gypsum);
- hydrolysis – feldspar changes into clay by reacting with the H^+ and OH^- ions in water;
- solution – minerals are dissolved by water;
- carbonation – weakly acidic water dissolves limestone.

Factors affecting type and rate of weathering

Some kinds of rock are more resistant than others, depending on their mineral composition and hardness. Igneous and metamorphic rocks tend to be more resistant than sedimentary.

Dark rocks absorb more heat so exfoliate more rapidly. Coarse-grained rock and rock with many joints and bedding planes allows greater water penetration.

Each weathering process has ideal climatic conditions (temperature and precipitation):

- Freeze-thaw needs seasonal or diurnal fluctuations around 0°C and available water.
- Chemical processes are all accelerated by high temperatures and so these are effective in tropical regions.

Decide the type of climate, (temperature and precipitation) which best suits each type of weathering

Consider how precipitation is particularly important for chemical weathering.

An idea of the amount of chemical weathering taking place in an upland area could be determined by measuring the concentration of solution load in a stream with a conductivity meter.

4 Weathered rock moves downslope, in response to gravity, in mass movements. Slope stability is the ability of the slope to maintain itself at a given angle.

The point at which gravity exceeds friction is the point of failure and shearing results. An unstable slope has a high mass of material, a steep angle, and a high water content. Water lubricates the rock debris, forces soil grains apart and adds extra mass.

Mass movements can be rapid (such as falls, slides and slumps), less rapid flows, or very slow soil creep. This last mass movement is caused by rainsplash erosion of individual particles, frost heaving of particles downslope and the expansion and contraction of soil.

Write a short list of examples for each type of mass movement, e.g. rapid flow: Aberfan disaster, 1966.

Falls, slides and slumps

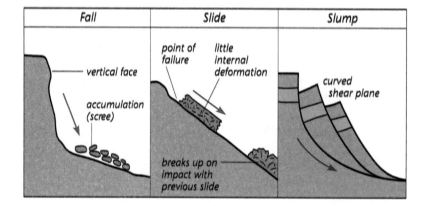

Tectonics and Surface Processes

Limestone landscapes

The various features of limestone landscapes are sometimes called 'karst scenery'.

In some regions such as the Yorkshire Dales, areas of limestone are found. Limestone is made up of millions of tiny fossils of corals and other small invertebrates. This shows that it was formed on the bed of a tropical ocean.

This rock is hard and white. As the rock is permeable, there are no surface streams. Limestone is not porous and so cannot soak up water but it allows water to pass through numerous minor structures such as vertical joints and horizontal bedding planes which let in water.

Limestone can dissolve slowly in water by a chemical weathering process called solution. Rain absorbs carbon dioxide from the air, forming a weak acid which changes the calcium carbonate in the limestone into calcium bicarbonate, which is washed away in streams.

This weathering produces certain distinctive features:

FEATURE	DESCRIPTION
Limestone pavements	On flat surfaces cracks or joints are enlarged by solution. The blocks between them, which form the pavement, are called clints.
Scars	Steep limestone cliffs
Springs	Underground streams emerge at the surface when permeable limestone meets the impermeable rock underneath.
Swallow holes	A large joint down which streams disappear
Gorges and dry valleys	When it was much colder, the ground was frozen and rivers could run on the surface.

Draw simple sketch diagrams for each. They do not have to be elaborate, but should include two or three labels.

Tectonics and Surface Processes

45 minutes

Use your knowledge

1 Why do volcanoes often occur on both constructive and destructive plate margins? Refer to specific locations and examples.

2 Describe the initial and long term effects of an earthquake or volcano that you have studied. Suggest ways in which the effects of such events on people can be minimised.

3 Describe and explain the characteristics of weathering in tropical regions.

4 a) Define the term 'weathering'.
b) Which weathering process would be most important in a region with high mean temperatures and very low rainfall, such as the hot deserts?
c) Under what climatic conditions does freeze-thaw action occur most intensively?
d) Define the term 'mass movement'.
e) Describe the ways in which humans can control or reduce mass movements on slopes.

Hint

You will be expected to name examples or describe case studies for most questions.

Your answer can be divided into the three obvious parts.

Consider how climate affects the various weathering processes.

If you can, give examples of places where such measures have been taken.

15 minutes

Test your knowledge

1 What is a glacier?

2 What are the inputs to and outputs from the glacial system?

3 Explain the concept of the pressure melting point.

4 What is medial moraine?

Answers

1 A glacier is a large mass of ice which slowly moves down a valley due to gravity. The huge mass of accumulated snow is compressed into ice. Under this pressure, ice at the base can melt and so allow the glacier to move. The surface is often cut by large cracks or 'crevasses'. **2** Inputs are precipitation (snowfall) and avalanches onto the glacier from above. Outputs are meltwater streams, evaporation directly from the glacier surface and 'calving', when it reaches the sea or a lake. **3** Under normal atmospheric pressure, ice changes to water at 0°. Under higher pressure, such as at the base of a thick glacier, ice may melt at lower temperatures. **4** The lateral moraines of two glaciers merge when they meet, leaving moraine along the centre of the valley.

 If you got them all right, skip to page 37

Glacial Environments

Improve your knowledge

In colder climates, either during previous ice ages or in mountainous polar regions, precipitation falls more often as snow. Cool, short summers mean that less ice and snow will melt. Snow becomes compressed by the weight of other subsequent snowfalls on top and so changes to a dense snow called firn. As more air is slowly pressed out, firn gradually changes into ice.

The recent ice ages occurred during the Pleistocene era. Learn a few terms such as this to use in your answers.

1 A glacier is a large tongue-shaped mass of ice, which slowly moves down a valley due to gravity. The huge weight of accumulated snow is compressed into ice. Under such pressure, ice at the base melts and so allows the glacier to move. There is usually a meltwater stream, which runs continually in summer from the foot of the glacier. The surface is often uneven and cut by large cracks or crevasses.

2 The glacial budget is the net balance between accumulation (snowfall added to the glacier) and ablation or melting.

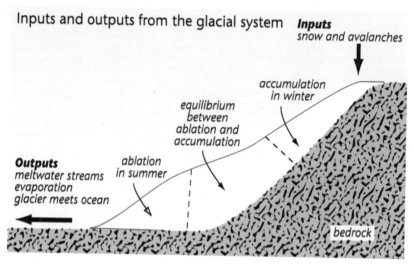

Inputs and outputs from the glacial system

Inputs snow and avalanches

accumulation in winter

equilibrium between ablation and accumulation

Outputs meltwater streams evaporation glacier meets ocean

ablation in summer

bedrock

If the question permits, you can get good marks by answering with a fully labelled diagram.

In the short summers, ablation will exceed accumulation and the mass and size of the glacier will decrease, whereas the opposite is true in winter. In temperate mountainous areas, such as the Alps, summer ablation and winter accumulation are almost the same and so the glacier is in equilibrium.

3 Ice movement

At the surface of the glacier, the ice melts at 0°C, as we would expect. However, within the glacier and especially at its base, ice can melt at lower temperatures, perhaps -2 or -3°C. This occurs due to intense pressure from the weight of ice above. The ice is said to have a lower pressure melting point. Glaciers in temperate regions would therefore have more melting of ice at the base and it is this 'lubrication' which allows the glacier to move. Consequently, temperate glaciers have faster rates of movement than polar glaciers.

Speed of movement is increased by:

- steep gradients;
- high accumulation from plentiful snowfall;
- small glaciers responding quickly to rising temperatures;
- relatively high summer temperatures.

Processes of glacial erosion:

Process	Description
Freeze-thaw or frost shattering	This is most effective in areas with a summer thaw. Water enters cracks in the rock, freezes and expands, widening cracks and producing a lot of rock debris. This falls into the glacier and becomes embedded in it.
Plucking	Water freezes onto and around rock, then pulls away small and large fragments of rock as the glacier moves away.
Abrasion	Debris from freeze-thaw or 'plucked' from the bedrock can cut and scour the bedrock, forming smoothed rock features.
Rotational movement	As the glacier moves down the valley, there is often a rotational movement about a point. This allows for very effective erosion, so deepening the valley floor.

Making tables like this for your other revision notes can help you to learn them more effectively.

4 Glacial features

Corries: sometimes called cirques, these are armchair-shaped hollows found in mountainous areas (Scotland, North Wales and Cumbria in the UK).

They have a rounded rock basin and a steep backwall and are found in areas which once had a glacial climate such as the UK in previous ice ages.

Snow collects in hollows on colder, north-facing slopes. Processes of glacial erosion enlarge the hollow, processes of weathering erode the backwall and a rock lip is formed on the edge of the corrie. When the ice has retreated, a deep, round tarn or lake is often left behind.

Formation of a corrie

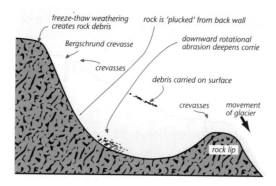

You should also be familiar with the processes of formation of the following glacial landforms:

- arêtes and pyramidal peaks (steep ridges between two or more glaciated valleys);
- glacial troughs (U-shaped valleys);
- truncated spurs and hanging valleys (tips of pre-glacial interlocking spurs eroded by glacier to leave cliff-like features);
- roche moutonné (mass of resistant rock with smoothed up-valley side and jagged, 'plucked' lee side).

Draw a simple labelled diagram and write a paragraph explaining the formation of each.

Moraine is a landscape feature which is formed when the rock material carried by a glacier is deposited.

Type of glacial moraine	Description
Lateral moraine	Moves along the side of the glacier. Comes from frost shattering of valley sides. Leaves low banks along valley sides.
Medial moraine	The lateral moraines of two glaciers merge when they meet, leaving moraine along the centre of the valley.
Terminal moraine	Material pushed in front of the glacier forms a large mound across the valley.

Glacial Environments

Use your knowledge

 a) Define and explain the term 'glacial budget'.
   b) Explain how: decrease in temperature and decrease in precipitation influence the glacial budget.

 a) Explain how processes of weathering and erosion may result in the formation of a corrie or cirque.
   b) Suggest how the characteristics of the corrie may have changed since the glacial period.
   c) What evidence can be used to deduce the maximum extent of the ice flow in a valley?

Hint

You could draw a simple graph to show the annual pattern.

Explain the effects step by step.

Say how one can increase the effectiveness of the other.

What process (es) will continue after the glacial period?

Consider glacial deposition.

15
minutes

Test your knowledge

1 Name the atmospheric system that transfers heat from the warm equator to the colder north and south poles.

2 What is relative humidity?

3 What is the difference between environmental and adiabatic lapse rates?

4 If rising air cools more rapidly than that surrounding it, a state of _____ exists. If the rising air is warmer, it continues to rise, causing _____.

5 Describe how temperatures may differ in an urban area.

Answers

1 the heat budget 2 RH is the amount of water held in the air, as a % of what would be held if the air was saturated. 3 ELR is the natural rate of temperature decrease with altitude. ALR is the rate of temperature decrease due to lower pressure at high altitudes. 4 stability, instability 5 An urban heat island' exists whereby if winds are low, temperatures can be up to 5°C higher in the built-up city centre.

 If you got them all right, skip to page 42

Meteorology and Climate

45 minutes

Improve your knowledge

1 The earth's temperature stays relatively constant. The insolation (incoming short-wave radiation) it receives and the outgoing radiation must be balanced.

Insolation is lower in higher latitudes, in winter and at night, depending on the earth's position and tilt. Ozone, CO_2 and water vapour reflect or absorb some. The rest (24%) is scattered as it reaches the earth. Outgoing long-wave terrestrial radiation is blocked by CO_2.

Although the insolation received varies with the length of night and day and the warm and cold seasons, the following factors are also important:

- Oceanic regions such as the Pacific can hold more heat energy than land surfaces and can store this heat in winter.
- Winds can blow from warm to colder areas, thus warming the latter up.
- Aspect is the direction in which a hillside faces. South-facing slopes are warmer than those facing north.

The deficit of heat at the north and south poles and surplus in tropical regions is balanced by the heat budget.

Heat is transferred by:

Horizontal transfer	Vertical transfer
ocean currents	convection
jet stream winds	conduction
hurricanes	radiation
depressions	transfer of latent heat

2 Humidity is a measure of the amount of water held in the atmosphere.

Absolute humidity is measured in grams of water vapour per m³ of air. The air can hold more water at higher temperatures.

Relative humidity (RH) measures the amount of water in the air, as a percentage of what would be held if the air was saturated (100%). On a very humid day the RH may be 80%, but a desert may only have 15%.

You must learn the meaning of all the different terms used for the various energy transfers in this chapter.

You should be able to write full paragraphs to explain processes such as these.

Humidity is measured in a weather station using a wet-and-dry-bulb hygrometer. This has two thermometers, one of which is kept moist. The water evaporates, causing cooling. The difference between the two can be converted to relative humidity.

As unsaturated air cools, it becomes saturated when it reaches the dew point. Further cooling causes condensation into either droplets of rain or ice crystals.

In just the same way, warm air inside a room condenses to form tiny droplets on cold windows.

3 Temperature decreases with altitude. This rate of change is called the environmental lapse rate (ELR).

A reduction in pressure also causes air temperature to fall, so when air rises, its temperature will fall due to a fall in pressure. This rate of change is called the adiabatic lapse rate (ALR).

Study these diagrams carefully and slowly as you read the explanation. Which two variables do the two axes represent? How do these variables change in relation to each other?

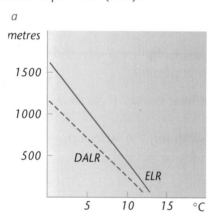

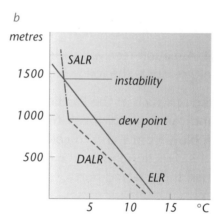

If the rising air is unsaturated, its temperature falls at a fixed, dry ALR (DALR) of 9.8°C per 1000m.

If dew point is reached and condensation occurs because of this rising and cooling, the air may continue to rise but whilst doing so it will cool at a slower and variable saturated ALR (SALR) of between 4° and 9° per 1000m.

4 Up to 1000m, the air rising at the DALR is still cooler than the surrounding air (ELR) and so is stable as in diagram (a), above.

In diagram (b), above, dew point is reached at 1000m and the air cools more slowly at the SALR. From 1500m, the air is warmer than the surrounding air which is cooling at the ELR and so instability results in cumulus cloud and showers.

Droplets form in the cloud when water vapour condenses around dust particles. When the temperature is below zero, ice crystals will from, giving hail if they coalesce or snow if they cluster loosely in calm conditions.

Rain is formed by the growth of ice crystals (by sublimation) which melt as they fall (Bergeron-Findeisen process), or by the coalescence of droplets.

5 Large urban areas may have different annual patterns of temperature and precipitation. They have altered climates due to the following reasons:

- Homes and industry generate heat.
- Homes and industry add more water to the atmosphere.
- Cities create dust (condensation nuclei) which helps the formation of rain droplets.
- Tall buildings alter winds and airflows.
- Large urban areas have a different albedo or capacity to reflect heat.

Climate feature	Changes	Explanation
Temperature	An 'urban heat island' exists whereby if winds are low, temperatures can be up to 4 or 5°C higher in the built-up city centre.	Buildings absorb heat during the day and release it slowly at night. Homes, industry and cars release heat.
Precipitation	Cities may receive up to 15% more rain and snow than surrounding rural areas. Frequent occurrence of fog.	Urban heat can create strong 'thermals' which may cause heavy rainstorms. Precipitation as fog is common due to abundance of dust particles.
Wind	Overall reduction in mean wind speeds but occasional 'funnelling' between tall buildings.	Houses and factories act as windbreaks reducing mean wind velocity.
Humidity	Lower relative humidity in cities.	Little vegetation so low evapotranspiration. Air can hold more water as it is warmer.

The falling temperatures towards the outskirts of an urban area are often shown by isotherms in concentric rings around the city.

45
minutes

Use your knowledge

Hint

1 Study the diagram below:

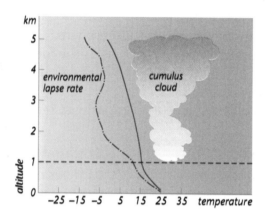

a) Label the following:
 – condensation level
 – saturated adiabatic lapse rate
 – dry adiabatic lapse rate.
b) Describe the climatic conditions usually produced by unstable air conditions.

Remember that
clouds form in
saturated air.

2 a) Explain the following terms:
 (i)   sensible heat transfer
 (ii) use of latent heat in evapotranspiration.
b) How does higher cloud cover affect the following?
 (i)   solar radiation received
 (ii) solar radiation reflected
 (iii) long wave radiation – outgoing
 (iv) heat absorbed in ground
c) Explain why a car park would have a lower surface temperature at night than an area of open parkland.

You need a short
but detailed
paragraph for each.

Consider the
factors in part b for
your answer.

10 minutes

Test your knowledge

1. What is the difference between population distribution and population density?

2. Why are some places sparsely populated whilst others are densely populated?

3. Define fertility rate.

4. Briefly describe population changes at each stage of the demographic transition model.

5. What are the checks in Malthus' theory?

Answers

1 Distribution refers to the way people are spread out within an area, density refers to the number of people per unit area. Density maps often mask distribution within a country. 2 There are physical (climate, relief), economic (resources), social (culture) and political (government policy) factors which affect where people live. 3 Fertility rate is the number of live births per woman of reproductive age (15–50); if fertility is 2.1 children then it is likely that a population will replace itself. This measure is more accurate than crude birth rate. 4 Four stages: high fluctuating, early expanding, late expanding, low fluctuating. Possible 5th stage: declining. 5 Preventive – human decisions to reduce births; positive – events that decrease population such as war or famine.

 If you got them all right, skip to page 48

30 minutes

Improve your knowledge

1 World population

Six billion people now inhabit the Earth. Population distribution is uneven; less than 10% of the total population live in the southern hemisphere, 50% live between 20°N and 40°N and less than 0.5% live north of 60°.

Europe is the most densely populated continent and Australasia the most sparsely populated. However, only 15% of the world's people live in Europe compared with 59% in Asia. Asia has some of the densest regions in the world – Eastern China and coastal areas of Japan – yet large areas of these countries are hostile and remain virtually uninhabited.

Distribution is often shown using a dot map, whereas density is shown using a choropleth map

Map of world population distribution

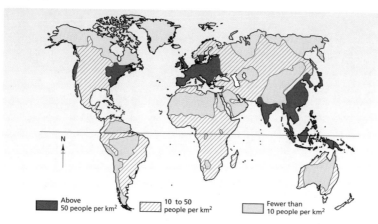

N

Above 50 people per km² — 10 to 50 people per km² — Fewer than 10 people per km²

2 Some factors affecting population distribution and density:

- temperature; reliability and amount of precipitation; humidity; associated diseases and pests;
- relief and altitude; effect on communications;
- farming: soil fertility, length of growing season;
- water supply and rivers/ports for transport;
- mineral resources to develop industry;
- government policy and investment;
- development of urban areas with opportunities.

Give examples of places that illustrate these factors. Are the places sparse or dense?

Population

Population structure

Key terms:

Learn the birth rate, death rate, fertility rate and infant mortality rate for your chosen LEDC and MEDC.

- Crude birth rate – number of births per thousand of the population per year; crude because it includes all people not just the fertile population (women aged 15–50).
- Death rate – number of deaths per thousand per year; in development indices, measures of infant mortality are more significant (the number of deaths of live child births aged under one year per thousand per year).
- Life expectancy – the average number of years from birth that a person can expect to live.
- Natural increase – the difference between birth rates and death rates, which along with net migration indicates the growth of a population. Percentage population growth can then be calculated; 4% population growth per year is a considerable amount.

Factors affecting birth rates include:

- religious beliefs and cultural tradition;
- whether children are seen as an economic asset and as insurance for old age or as a burden;
- the status of women; whether they are educated and work in paid employment;
- whether people live in rural or urban areas;
- the state of economic development within a country;
- government policies (e.g. China's one child policy).

Factors affecting death rates and life expectancy include:

Apply these factors to your chosen LEDC and MEDC.

- access to medical care and immunisations;
- sanitation and access to clean drinking water;
- success of agriculture and food distribution;
- calorie intake and diet;
- educational levels giving access to employment;
- whether people live in rural or urban areas;
- natural disasters and the resulting outbreaks of disease;
- working conditions and safety in factories;
- political climate, e.g. war or structural adjustment policies.

4 Population pyramids and the demographic transition model

An age–sex pyramid shows the percentage of males and females of each age group within a population. The pyramids can be used to predict future changes in population growth. The demographic transition model shows changes over time in birth rate, death rate and total population of a country.

The demographic transition model with pyramids for each of the four stages

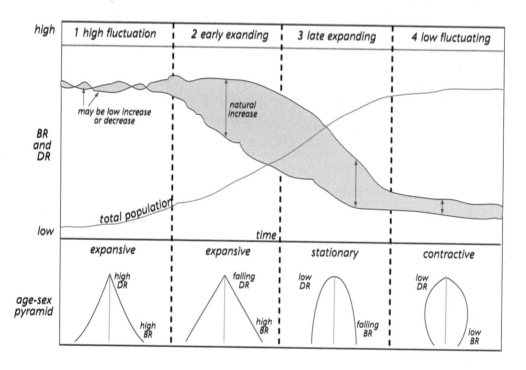

A dependency ratio expresses the number of young or old dependants who rely on each economically active adult. LEDCs tend to have a high proportion of young dependants, whilst MEDCs tend to have an old age structure.

You should be able to annotate the population pyramid for your chosen LEDC and MEDC and show which stage it links to in the demographic transition model.

Remember that the ages of the economically active population will vary between countries.

Population

5 Theories of population growth

In the eighteenth century, Thomas Malthus suggested that population would outstrip resources because population grows geometrically (1, 2, 4, 8, 16, 32) whilst food supply grows arithmetically (4, 5, 6, 7, 8, 9); overpopulation would result.

He suggested that there were two checks to population growth:

- preventive (negative): human methods of reducing population growth, such as delaying marriage;
- positive: events such as famine, disease and war which would reduce population size.

Malthus believed that there was an optimum population size, where the carrying capacity would not be exceeded.

Boserup theorised that as population increased, food output would increase through technological innovation

Governments may implement population policies in response to under- or over-population. China's one-child policy (now relaxed) is well documented. Although harsh, it demonstrates the desire of the government to attempt to reduce future demographic problems. The policy was most successful in urban areas, where propaganda was more easily distributed and controls were stricter. The negative side effects include a generation of children with no siblings, and the female infanticide that sometimes resulted from parents' desire to have a son instead of a daughter.

You may be asked to discuss the success of a population policy

45 minutes

Use your knowledge

1. Examine the factors influencing population distribution in a named region.

2. Describe how population characteristics have changed in a named country over time.

3. What is overpopulation?

4. Explain why some areas of the world could be described as overpopulated.

5. What effects can overpopulation have on the physical environment?

Hint

Give an example and a range of physical and human factors

Think about total population, BR, DR and the demographic transition model

(3) Define the term, relate it to theory and give an example

Think of examples in different parts of the world

10 minutes

Test your knowledge

1 What is the difference between forced and free migration?

2 Describe the factors that might contribute to a person's decision to migrate.

3 What is distance decay?

4 List five benefits for the host country of international migration.

5 How might governments try to discourage internal migration?

Answers

1 In a forced migration the migrants have no choice whether they move (such as refugees) whereas free migrations are voluntary. **2** Push factors that make them want to leave their home and pull factors that attract them to a new place. **3** As distance from home increases, the number of migrants decreases due to the 'friction' of distance. **4** Workers have new skills; workers can fill less desirable jobs at lower wages; labour shortages may be overcome; new culture is brought to a country; migrants may be entrepreneurial. **5** By improving the areas that people are leaving and therefore reducing the strength of the push factors (e.g. by developing new industries to increase job opportunities, by improving housing standards and education). In LEDCs this may be the more peripheral urban areas or rural areas, in MEDCs this may be the inner areas of large conurbations.

✔ **If you got them all right, skip to page 53**

30 minutes

 Improve your knowledge

1 Classification of migration

Migration may be classified by:

- choice – forced or free
- distance – internal (within a city, region or country) or international (between countries or continents)
- length of stay – daily (commuting), seasonal, semi-permanent (for several years) or permanent (once in a lifetime).

Study a variety of forced and free migrations at different scales so you can apply them to questions.

Migration affects the population distribution, population totals and population structure within a city, region or country. The migration balance is the difference between the number of immigrants (people who enter a country) and emigrants (people who leave a country).

Some examples of international migrations

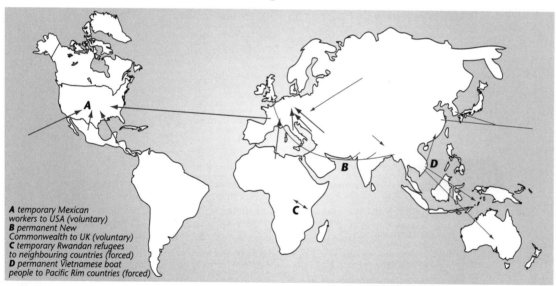

A temporary Mexican workers to USA (voluntary)
B permanent New Commonwealth to UK (voluntary)
C temporary Rwandan refugees to neighbouring countries (forced)
D permanent Vietnamese boat people to Pacific Rim countries (forced)

2 Causes of migration

Forced migration (push factors):

- war
- religious or political persecution

Migration

- environmental disaster or natural hazards
- slavery or forced labour
- resettlement or redevelopment

Voluntary migration (push and pull factors)

- economic opportunities (jobs, wages)
- climate and environmental factors
- new land or towns being developed
- educational and medical amenities
- quality of housing and infrastructure

3 Models and laws of migration

Models of migration have been developed in an attempt to explain or classify types of migration.

Gravity models show the degree of interaction between two places. The volume of migration can be calculated by using a formula. Larger places will attract more migrants and as distance from the place increases so the level of migration will decrease (distance decay).

Ravenstein's laws of migration identify the major characteristics of migration and have been updated to show more recent trends.

You will be given formulae to use and substitute values into. You need to be able to explain the answers they generate.

A modern approach to Ravenstein's laws of migration:

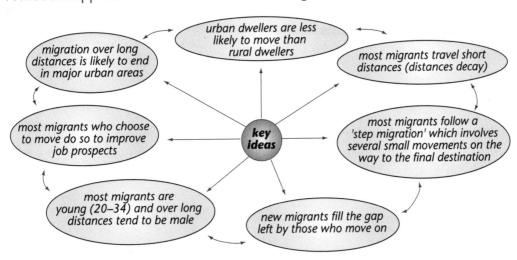

51

Benefits and problems of migration

Both the regions/countries of origin and destination can benefit but can also lose out as a result of migration.

If the movement is within the country, it can often be from peripheral to core regions. This results in a drain of skilled young labour to the most industrialised and prosperous areas. This can help to fuel a process of cumulative causation whereby the core regions become an even greater focus of wealth and employment. However, too much migration to the most attractive regions too quickly can have undesirable effects such as housing shortages or fast rises in the cost of living.

Make sure you know a case study of internal migration. Brazil and the UK are well documented.

For host countries immigrants can provide labour and skills. However, earnings are often sent home as remittances and ethnic tensions can be problematic, especially during periods of economic downturn. Migrants are often young males and often leave behind an unbalanced population structure. A loss of trained or educated people, especially from LEDCs (less economically developed countries) to MEDCs (more economically developed countries), can be of real detriment to the country of origin. If migrants return to their country of origin they can take with them new skills and wealth.

Examples include Turks to Germany or West Indians to the UK.

Refugees can be a huge drain on the host country, particularly if the movement is on a large scale, such as the 2.4 million Rwandan refugees who fled to neighbouring countries in 1994.

 Government migration policies

- All countries have barriers to international migration which may be selective according to the country of origin. Most countries have tightened their entry conditions in recent years, as huge influxes of people can be detrimental to economies.
- Internal migration is difficult to stop. Some countries have organised resettlement schemes while others give regional aid to increase the appeal of certain areas.

Migration

Use your knowledge

Hint

You will not have seen many of the models or diagrams you are given in exam questions, but apply your knowledge and skills of interpretation to them.

1 Study the model of migration below.

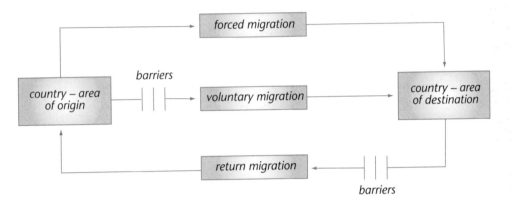

a) Discuss the types of barriers that prevent free migration.
b) Why might governments place barriers to migration?

2 With reference to examples, discuss the factors causing international migration.

3 What could be the environmental and social impacts of the arrival of large numbers of refugees? Refer to examples you have studied.

Factors and impacts can be human (social, cultural, economic, political) or physical.

Settlements and Urban Areas in MEDCs

Test your knowledge

1 What is the difference between site and situation?

2 What is a settlement hierarchy?

3 What is a central place?

4 Define the term 'urbanisation'.

5 What are the main differences between the Burgess model and Hoyt's model of urban land use?

6 List five factors that have contributed to inner city decline.

7 What term is used to describe the process by which people move from large urban areas to smaller towns and rural areas?

8 What settlements were built between the 1940s and 1970s to ease pressure on the large urban areas?

Answers

1 Site is the piece of land on which a settlement is built. Situation is the settlement's position relative to other human and physical features. **2** Settlement hierarchy is a list of settlements in order of importance and may be defined by population or number of functions. **3** Central places are those which exert a sphere of influence on their surrounding area because they provide goods and services. **4** Urbanisation is a process by whereby an increasing proportion of a country's population lives in towns and cities. **5** Burgess assumed cities grew in concentric rings from the CBD (Central Business District). Hoyt assumed city growth would be in wedges or sectors along transport routes. **6** loss of industries leading to unemployment, traffic congestion, air pollution, crime, poor quality housing etc. **7** counter-urbanisation **8** new towns, overspill towns and expanded towns

✔ **If you got them all right, skip to page 62**

45 minutes

Improve your knowledge

1 Site factors

The features of the actual place (site) at which a settlement is located influence its growth and development. Most settlements that exist today have their origins extending back several millennia; even new towns have historic cores. Often the name of a place gives a clue as to its site factors. For example, Oxford was built at a place where oxen could wade across the river at a shallow point in the river (ford). Cambridge was built at a bridging point on the River Cam.

Learn a case study of site factors such as your home town or a settlement in your region.

Key site factors include:

- wet point sites with access to fresh water from rivers, wells or springs (spring line villages on chalk escarpments);
- dry point sites with non-marshy land above a river's flood plain on a river terrace or bluff;
- flat land and shelter;
- defensive sites such as hill tops or in the loop of a river meander;
- resources such as building materials, minerals or fertile soils for farming;
- communications including locations at natural harbours or nodal points/route centres.

Practise looking for site factors on an OS map and drawing annotated sketch maps.

A successful settlement not only requires a good site but also a suitable situation relative to surrounding settlements, water sources, communication lines, resources and relief.

2 Settlement hierarchies

At the simplest level, settlements can be classified by size.

A settlement hierarchy:

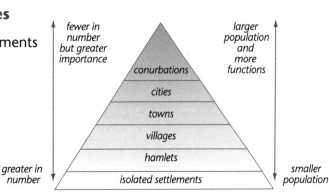

fewer in number but greater importance

larger population and more functions

conurbations
cities
towns
villages
hamlets
isolated settlements

greater in number

smaller population

Settlements and Urban Areas in MEDCs

Settlements can also be classified by function. However, it must be recognised that most towns and cities are multi-functional and that functions can change over time. In general the larger the settlement the more functions it has. Functions include markets, ports, industry, resorts, administration, services, education, housing and commerce.

Zipf's rank-size rule can be used to determine the relationship between the population of settlements in a country or region. It states that the size of settlements is inversely proportional to their rank. In other words the largest city is twice the size of the second, three times the size of the third and so on. When this is plotted on double logarithmic graph paper, the theoretical rank size rule shows a straight line.

Make sure you know how to plot graphs on semi-log and double log graph paper

Using the population of the largest city, the theoretical sizes can be calculated for the top ten ranked cities and this can be compared with the actual population size data.

The rank size rule

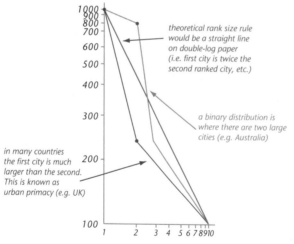

theoretical rank size rule would be a straight line on double-log paper (i.e. first city is twice the second ranked city, etc.)

a binary distribution is where there are two large cities (e.g. Australia)

in many countries the first city is much larger than the second. This is known as urban primacy (e.g. UK)

3 Central places and settlement patterns

Individual settlements have different shapes or morphologies, including nucleated, dispersed, cruciform and linear.

Central places are at the key to Christaller's theory that tries to explain why a certain pattern and hierarchy of settlements develops in an area. All settlements have a sphere of influence that is determined by the number and type of functions. To support a particular service or function a certain size of threshold

The sphere of influence can be worked out theoretically using Reilly's breakpoint formula or in the field by asking people where they have travelled from to use certain services.

population must live within range. Therefore larger settlements contain high order functions whilst small settlements mainly offer low order functions.

Settlement patterns can be investigated using 'nearest neighbour analysis' to determine whether their distribution is clustered, random or uniform.

4 Urbanisation

Nearly half the world's population live in towns and cities. In 1990, about 70% of the population of MEDCs lived in urban areas, with many countries such as the UK having a figure of around 90%. Urbanisation occurred rapidly during the Industrial Revolution, as unemployed rural labourers migrated to the expanding towns to find factory work and jobs in mining. Urban areas grew even more rapidly because of high natural increase and higher life expectancies associated with improvements in sanitation and medical care.

5 Urban land use

As urban areas have grown, geographers have noticed various spatial patterns and have used these to devise models to help them understand urban growth. However, it must be remembered that whilst many cities do exhibit similarities, they also have distinct layouts that can be explained by local factors.

Urban land use models

Burgess: concentric zone model (based on Chicago)

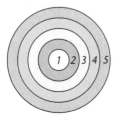

1 CBD; **2** zone of transition (inner city); **3** zone of working men's homes; **4** residential zone; **5** commuter zone

basic assumptions cities grow outwards from the centre and different socio-economic groups inhabit different zones

criticisms assumes a uniform plain and does not mention transport links or industry

Hoyt: sector model (based on 142 American cities)

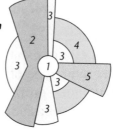

1 CBD; **2** wholesale, light manufacturing **3** low-class residential; **4** medium-class residential; **5** high class residential

basic assumptions housing patterns in cities were determined by ability to pay high rents, often along major roads

criticisms does not include redevelopments or modern edge-of-city schemes

Harris–Ullman (multiple nuclei model)

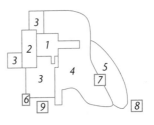

1 CBD; **2** wholesale, light manufacturing **3** low-class residential; **4** medium-class residential; **5** high class residential; **6** heavy manufacturing; **7** outlying business district; **8** residential suburb; **9** industrial suburb

basic assumptions land use patterns in cities are not built around a single centre but around a number of different nuclei which may be old or more recent

criticisms each zone tends to have more than one land use. This model is more complex and therefore becomes prescriptive not predictive

6 Land use zones within cities

Although the models all suggest different reasons for the spatial pattern found in cities, certain functional zones can be identified that are common to most urban areas.

A central business district (CBD) contains the principal commercial streets and major public buildings and is the centre for retail, commerce and administration.

A CBD has the highest land values because it is the most accessible part of the urban area, containing transport termini and the route focus of roads

The CBD is dynamic. Some zones are assimilated (by redevelopment or conversion of buildings); other parts are discarded as they become derelict or more run down. The CBD tends to be an area of constant redevelopment and renewal.

Many CBDs are facing increasing competition from out-of-town developments such as retail, leisure and business parks which are located in the suburbs next to ring road or motorway junctions. Small CBDs have suffered more – one obvious change is in retailing, where high order stores have closed to be replaced by discount stores and charity shops.

Traditionally the inner city (or twilight zone) refers to the nineteenth century residential and industrial areas occupied by terraced housing and factories.

Geographically, the inner city is inexact and are not necessarily central in location, in fact some outer city council estates exhibit similar socio-economic characteristics and may suffer from the same type of multiple deprivation (high unemployment, overcrowding, environmental problems) as inner cities. Both situations lead to population loss and a concentration of problems.

Since the 1960s, inner cities have been the focus of much redevelopment, since they have tended to be the areas which have suffered from inner city decline as a result of the closure of manufacturing industries.

Do a field work study to investigate the land use in an urban area; identify the CBD and the parts that are improving and the parts that are being abandoned.

Use census data to find out some of the socio-economic indicators in different parts of an urban area.

Settlements and Urban Areas in MEDCs

A downward spiral of inner city decline

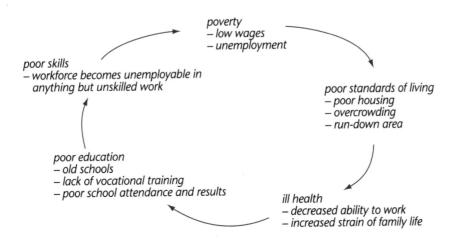

poverty
– low wages
– unemployment

poor skills
– workforce becomes unemployable in
anything but unskilled work

poor standards of living
– poor housing
– overcrowding
– run-down area

poor education
– old schools
– lack of vocational training
– poor school attendance and results

ill health
– decreased ability to work
– increased strain of family life

Urban policy in the inner cities

In the 1960s the policy focus was urban redevelopment and renewal in which old slum areas were demolished and replaced by high-rise blocks. However, there have been many subsequent problems with these developments (including poor design and damp problems, unsuitability for young families or the elderly as lifts do not work, youths leaving school with no qualifications, crime and drugs issues related to unemployment).

From the 1980s the focus has been more related to urban rehabilitation, whereby the existing fabric of the city is improved, rather than whole scale demolition. Some renewal has been successful, but this can often lead to gentrification because as the area improves, house prices increase and more wealthy people move in, displacing the original inhabitants.

A case study of change is in Manchester's Hulme district, where the notorious high rise 'Crescents' built in the 1960s exhibited many symptoms of social malaise. Hulme City Challenge initiated in the early 1990s has resulted in the rebuilding of Hulme with low-rise family housing and apartments. Public, private and European Union funding has been used in this successful scheme, Hulme is now experiencing a net in-migration for the first time in many decades and house prices are rising.

Make sure you know a case study that highlights some of the successes and problems of urban policy.

Other government policies have included Enterprise Zones and Urban Development Corporations.

Settlements and Urban Areas in MEDCs

Suburbs

Suburbs are the result of the outward expansion of urban areas and contain mainly medium- to low-density residential areas and other functions requiring more space.

Much of the housing dates from the 1920s to the present day and the outward growth is a response to improved transport links and personal mobility (the car). Outward urban growth along main roads is sometimes referred to as urban sprawl.

Many traditionally inner urban functions have been moving to suburban locations due to lower land values and the availability of greenfield sites.

Counter-urbanisation was first noticed in the USA in the 1970s, but has spread to most industrialised nations. It does not mean that everyone is moving to the countryside. However, people are migrating away from the large conurbations to much smaller settlements within commuting distance of the larger urban areas.

The explanation is based on the fact that people no longer like to live in large cities because the quality of life is lower, due to more congestion, pollution, crime, vandalism and noise. It tends to be better-off families with young children and retired people who move away from the cities. Lower-income groups including more recent immigrants and the unemployed remain and are concentrated in the inner city areas.

Census data will show the decline in major city populations since 1961. The census is conducted every 10 years; the next one will be in 2001.

Population loss from the large conurbations has made the problems more acute for those who remain, as public services are cut and council revenue decreases. However, there has been some reversal of the trend as up-and-coming cities such as Manchester and Leeds are attracting professional workers into the gentrified areas (converted warehouses and mills) on the edge of the CBD.

7 Urban planning

Various strategies have been used by successive governments to control and contain urban growth.

Greenbelts were established in 1947 to curb urban sprawl by imposing planning restrictions on the areas around the UK's major cities. Unfortunately, this has lead to residential and industrial development 'leap-frogging' the belts and expanding settlements beyond the greenbelt. In the South East there is great pressure to permit building on the greenbelt land as demand for housing continues to increase because of:

- counter-urbanisation;
- the continued economic growth of the South East due to the growth of hi-tech industries and financial services;
- demographic changes, in particular an increase in the number of single people living alone who wish to buy their own properties.

New towns were built from 1946 to:

- relieve overcrowding in the major conurbations;
- relocate the overspill population from inner city slum clearance schemes;
- attract people to modern housing in a semi-rural environment;
- act as growth poles in areas of high unemployment.

New towns were designed to be self-contained. However, some of the early ones were built too near London and were quickly engulfed by urban sprawl. The success of new towns has been varied. Washington in the North East has attracted many foreign firms. However, Skelmersdale, near Liverpool, has high unemployment and poor urban services.

Overspill and expanded towns were also established to accommodate population from the cities; however, these were based on existing settlements.

New towns may continue to have a different age structure from other urban areas as the children of the original inhabitants now have families. Retirement towns also have different age structures.

Settlements and Urban Areas in MEDCs

45 minutes

Use your knowledge

1. Draw an annotated sketch map to show the site factors influencing the development of a named settlement.

2. How does the concept of central places help to explain the pattern of settlements in an area you have studied?

3. How relevant are land use models in explaining land use in towns or cities you have studied?

4. Discuss some of the social, economic and environmental issues that may arise from the redevelopment of an inner city area. Refer to some examples that you have studied.

5. Argue the case either for or against the development of suburban retail parks.

Hint

Annotated means labelled. Maps are an important part of geography; use them wherever you can make them meaningful.

Think of positive and negative issues.

Think about their possible impact on greenbelts and the CBD.

10 minutes

Test your knowledge

1 How might a distinction be drawn between rural and urban areas?

2 What is the difference between commuter and suburbanised villages?

3 Why is rural depopulation occurring in remote villages?

4 What is the 'urban-rural fringe'?

Answers

1 Rural areas have settlements that are smaller in population and contain functions that are predominantly rural, for example agriculture. 2 Commuter villages contain people who do not work in the village but travel to nearby towns and cities. Suburbanised villages take on urban functions as their inhabitants demand amenities and services. Commuter villages may make the transition to suburbanised villages as their functions and population increases. 3 The more remote villages are suffering from population loss because they have declined; there is unemployment, poor housing and amenities, poor communications and a lack of services. Rural poverty is just as acute as urban poverty in many instances. 4 Urban-rural fringe is the edge of the town that merges into the countryside. It is an area that has a mix of urban and rural functions. Urban sprawl is when urban towns grow into the fringe area.

✔ If you got them all right, skip to page 67

Rural Areas in MEDCs

30 minutes

Improve your knowledge

1 Defining rural areas

Rural settlements may be classified by:

- size of population (they tend to be smaller in size, although this varies between countries);
- occupations of the resident population (traditionally these have been farming or cottage industries, but new footloose industries have been developed and many residents commute to work in urban areas);
- service provision (usually low order and limited);
- land use (less dense with more open spaces);
- social structure (an older age structure, especially in the more remote areas).

In India villages can contain up to 30,000 people, but are villages because the majority of residents are farmers.

2 Types of rural settlements

Our image of the English rural landscape may be of rectangular fields and hedgerows with interspersed farms, hamlets and villages. However, the countryside has undergone tremendous change since the 1940s.

In Britain 11% of the population live in rural areas. However, many more people live in areas that were once defined as rural, but have now taken on more urban functions as a result of socio-economic changes in the resident population.

Cloke identified four categories of rural area based on socio-economic data taken from the census. These ranged from extreme rural (such as central Wales) to extreme non-rural (suburbanised villages found around the main cities).

Dormitory settlements are those villages that have been taken over by commuters and retired people who work and shop in nearby towns. Often the village has few functions and the original population may face problems as the village shop is forced to close or bus services are cut.

Suburbanised villages are villages that have either been engulfed by large urban areas or exhibit urban functions and are therefore extremely non-rural in their character.

Consider the environmental impacts of cutting down hedgerows, for example loss of species and soil erosion.

Make sure you know an example of a village that had changed in character over the last 40 years.

Rural Areas in MEDCs

The influence of urban areas on their rural hinterlands

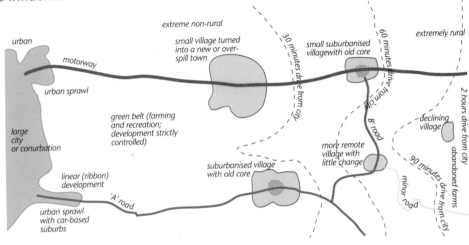

3 Rural development and change

Although it is usual to focus on the problems of inner cities, it must be appreciated that some of the most acute poverty occurs in rural areas, and depopulation is occurring rapidly. Some of the problems include poor housing lacking amenities, an ageing population structure, low paid jobs and unemployment, poor roads and poor facilities. Rural development strategies include the following:

- The Rural Development Commission is responsible for promoting rural industries and community development in England.
- The Mid-Wales Development Agency is responsible for promoting the economic and social development of 40% of the country. The agency offers loans, grants, rent exemptions and ready-built factories to attract new industries. Social development includes grants for community centres and improving communications in the remote areas.

Major changes in the rural landscape include:

- increased conservation including National Parks (the function of which is to preserve, enhance and promote rural areas of natural beauty), designation of Areas of Outstanding Natural Beauty (AONB) and Sites of Special Scientific Interest (SSSI);

Use census data to compare indicators of quality of life between rural and urban areas.

The Norfolk Broads is the most recent area to have been granted National Park status.

- a huge increase in the number of tourists visiting rural areas and using the amenities for leisure pursuits. This has caused conflicts with locals, especially at the most popular 'honey pot' sites;
- an increase in the ownership of second homes and holiday homes which leads to increasing rural property prices;
- the intensification of agriculture and the removal of hedgerows to create larger fields and amalgamate farms;
- the closure of many smaller farms due to falling incomes;
- changes in production including the introduction of new crops (such as oil seed rape), set-aside schemes, diversification into new activities and an increase in organic farming.

New manufacturing and extractive industries have opened in rural areas because of competing demand for land in urban locations. More attractive sites are available and demand for primary products such as limestone continues to rise. For example, the Toyota factory near Derby occupies a huge rural site and has been constructed to be sympathetic with the surrounding countryside.

A high rate of motorway and road building has consumed rural land and this has attracted local protests (e.g. at Twyford Down in Hampshire). The Manchester airport runway extension has caused much controversy.

Use local media sources and the Internet to identify local issues of conflict both in the countryside and on the rural urban fringe.

4 The urban-rural fringe

The urban-rural fringe (the area on the edge of cities) demonstrates the continuum between urban and rural areas. It represents an area where greenfield sites are under pressure to be developed for residential, retail and commercial purposes.

Despite lengthy planning disputes, out-of-town shopping centres – such as the Trafford Centre located on the M60 circular motorway east of Manchester and Bluewater in Kent – have been built. Although there has been a recent reversal of policy regarding edge-of-city developments, the character of rural areas has been irreversibly altered and encroached upon.

Rural Areas in MEDCs

Use your knowledge

Hint

1 Describe how the functions of a village within 30 minutes drive of a large urban area may have changed over the last 40 years.

The village may be dormitory or suburbanised.

2 Why have people moved out of large urban areas to smaller settlements in recent years?

3 a) How may villages in the more picturesque parts of the UK have become gentrified?
b) What are the possible social impacts of gentrification in such villages?

Apply your knowledge of urban gentrification to rural areas.

10 minutes

Test your knowledge

1 Why has urbanisation been so rapid in LEDC cities since the 1950s?

2 What is a 'primate city'?

3 How might the population structure be different in large cities from rural areas?

4 List five negative consequences of rapid urban growth.

5 Name a scheme that governments have implemented to improve squatter settlements.

6 How might urban land use differ between a typical developed and developing city?

7 How have some governments deliberately tried to direct growth away from the primate city?

Answers

1 Urbanisation has been a result of massive rural-urban migration and high natural increase within the cities. 2 Primate cities are the most important cities in a country, which attract the most people, resources and industry. 3 Population structure may be more youthful and male-dominated, as males are more migratory than females. 4 Lack of conventional housing leading to uncontrolled building of squatter settlements; urban sprawl as cities grow outward; poor living conditions lacking basic amenities such as sewerage systems and water supplies; unemployment and underemployment in informal jobs; congested transport systems that cannot cope with the ever-increasing population. 5 site and service scheme/self-help schemes. 6 The quality of housing tends to be reversed with the poorest housing on the edge of the city. 7 by building new towns and even new capital cities

 If you got them all right, skip to page 76

Settlements and Urban Areas in LEDCs

45 minutes

Improve your knowledge

1 Urban growth

Urbanisation has occurred at a very rapid rate in LEDCs (much faster than in MEDCs) since the 1950s as a result of:

- huge rural to urban migration caused by a combination of push and pull factors;
- high natural increase in the cities due to increased life expectancies, better living conditions, improved nutrition, immunisations and continued high birth rates.

Rural push factors include:

- inheritance laws creating fragmented farms too small to feed a family even at a subsistence level; landlessness;
- crop failure and natural disaster;
- unemployment due to mechanisation resulting in landless labourers;
- civil unrest and war.

Make sure you know a case study – migration from north east to south east Brazil is well documented.

Urban pull factors include:

- perception of a better standard of living through higher wages, improved healthcare and education;
- re-uniting with family members who have already migrated.

Increasingly there is a blur between rural and urban areas as farmland is swallowed up on the edge of cities and some migrants practice some aspects of rural life in the squatter settlements (such as crop cultivation or raising animals).

2 Millionaire and mega-cities

The unprecedented rate of urbanisation has resulted in most of the world's largest cities being located in LEDCs. Millionaire cities have more than one million inhabitants, whilst mega-cities have more than ten million. In 1990, LEDCs had a 40% urban population, with parts of South America more than 80% urban.

If you draw a scatter graph with a best fit line, there is a positive correlation between figures for urbanisation and GNP (gross national product).

The most highly urbanised LEDCs are those which have developed most in terms of their economies in recent years. Sub-Saharan African countries are the most rural and remain the least industrialised.

Distribution of the world's largest cities

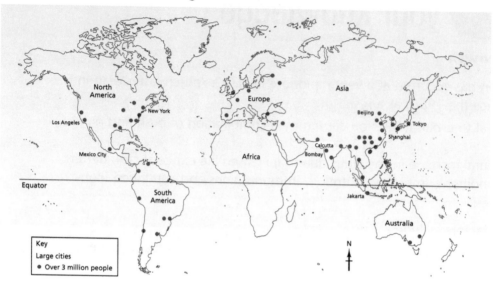

Many migrants have moved step by step to the largest city, often known as the primate city. Such cities are particularly characteristic of LEDCs where former colonial powers may have encouraged the growth of the capital city and economic development has not spread across the country. Certainly, many LEDCs exhibit urban primacy (which can be demonstrated using the rank size rule). These include Peru and Sri Lanka. Brazil has two dominant cities which concentrate development along the south eastern coastal strip.

Use a case study to show the effects of urban primacy – Brazil makes a good case study.

3 Urban age structures

Urban areas tend to have a fairly balanced but youthful population because:

- Young people are more migratory.
- Migrant families have many young children, although urban families are now likely to have a lower birth rate as women become more emancipated (with access to education, work and family planning).
- Both men and women tend to migrate because women are becoming increasingly educated and want to improve their lives.
- Employment aimed at women in hi-tech assembly jobs and textiles has become available as multinational firms set up in LEDCs. This is especially the case in the NICs (newly industrialising countries) such as Taiwan, Malaysia and Brazil.

Remember Ravenstein's laws of migration.

Settlements and Urban Areas in LEDCs

4 Urban problems

The rapid urban growth rates in many developing cities has, as might be expected, created a number of social, economic and environmental problems.

Housing is inadequate in quantity and quality to keep pace with the growing populations. In response, older housing has been subdivided into rooms, or storeys have been added to buildings. Rent is charged for such accommodation and many migrants find that they are so poorly or infrequently paid that they must build their own shacks. Despite the fact that they are often built illegally on council land, many shanty towns or squatter settlements have improved over time and offer satisfactory living conditions for many inhabitants. Councils may have bulldozed them in the past, but they now realise that adding infrastructure (running warer, electricity and drains) is cheaper than providing municipal housing. One of the more unusual informal settlements is the City of the Dead in Cairo where 2 million people live in a cemetery. In Hong Kong, squatters live on boats (junks) in the waterways and harbour.

Shanty towns have local names: favelas in Brazil, bustees in India, bidonvilles in North Africa.

Employment is not as readily available as migrants hope and many work in hazardous and arduous factory jobs. A vast number work in the informal economy where they create their own enterprises to earn money providing goods and services in the shanty towns. Many jobs are underpaid and people are underemployed. The vast numbers of people migrating with false hope for work is sometimes referred to as 'Dick Whittington Syndrome'.

Squatter settlements are often regarded as negative; beware of stereotypes and negativity when you write about them.

Services are under strain, meaning that many people do not have access to clean drinking water or proper toilet facilities. The streets can quickly become polluted with waste and sewage so diseases spread easily, especially when temperatures are high.

Hazards from flooding and landslides are also a problem because shanty towns are often built on marginal land such as slopes. Recently the shanty towns of Caracas suffered from severe flooding and landslides. Fires are also common because there are no building regulations.

Public transport is available but inadequate. Journey times are long due to overcrowding, congestion and the large distances between many of the peripheral areas and places of work. Cities such as Mexico City are very polluted as a result of fumes from vehicles and industries.

Education is probably better in the cities than rural areas and some people do succeed in breaking out of the poverty cycle. However, education is not free.

Healthcare is more accessible in urban areas, resulting in reduced infant mortality. This has helped to reduce birth rates in urban areas.

Social problems including crime, drugs, street children and prostitution are common.

Inequality is increasing despite the growing 'middle class' in many developing cities who enjoy high living standards of living and the comforts to which Westerners are accustomed. The gap between the rich and the poor continues to widen. In Rio de Janeiro 30% of the city's inhabitants live in shanty settlements. This rises to 45% in Mexico City and 60% in Calcutta.

5 Improving the conditions in cities

Governments have implemented a number of schemes aimed at improving conditions for urban dwellers.

Self-help schemes are aimed at the shanty town dwellers helping themselves to improve their lives. People are taught skills, advised on suitable building materials and encouraged to work together. After all, many shanty dwellers have already proved that they can be very resourceful by constructing their shacks from limited materials. Intermediate or appropriate technologies that focus on local expertise and materials may be used.

Site and service schemes are often complimentary to self-help schemes. The key issue is that the dwellings are made legal (people are given the land). The inhabitants then have more security and will be more inclined to invest in improvements. The authorities add infrastructure. In some of the best schemes the basic 'shell' of a house may be provided.

High rise flats have been built with varying success, but only solve the problem if the tenants can afford the rents. Otherwise rooms are sub-let or people are forced out of their homes and back to their former situation. Flats have been more successful in countries like Singapore and Hong Kong where industrialisation has accompanied urban growth.

In Lima (Peru) squatter settlements are called 'young towns' to emphasise the positive aspects.

Site and service schemes have been successful in Sao Paulo and Rio de Janeiro in Brazil.

It must be noted that most such improvement schemes are only small in scale and do not begin to tackle the volume of people or cause of the problem. Schemes are more likely to be successful if development is grass roots (involving local people) and affordable. In addition, most shanty dwellers live a very precarious existence; if the authorities decide to sell the land for more lucrative developments people will have to move on and compensation will be limited.

Involving the local people is a principle that applies to any scheme in an LEDC or an MEDC.

6 Land use in developing cities

Models have been developed to show land use in Latin American cities, which although useful should be treated as models and not reality.

Land use in a typical Brazilian city

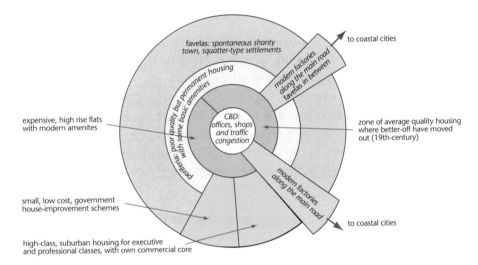

The CBD has a similar function to Western cities and may give the impression of a very wealthy city, being full of high rise blocks, shopping malls and apartments. Often space is limited and congestion acute.

The inner zones contain apartments owned by the middle classes and some subdivided dwellings where the better off working class may live.

Compare these trends with models of Western cities and other developing cities.

The outer zones contain squatter settlements, the oldest being nearest to the centre and the newest being at some distance from the CBD. Any marginal slopes or swampy areas will contain poor quality housing.

Industries tend to be along major transport routes especially main roads and railways.

7 Government initiatives to redirect development

LEDC governments attempt to redirect growth away from the primate city in order to reduce pressure and create a more even economic growth across the country.

New capital cities have been constructed by a number of governments across the developing world to create new growth poles and to make a break from their colonial past. One famous example is Brasilia – completed in 1960, it was built in a virtually uninhabited part of the country to house administrative functions, attract new industry and divert people away from the dominant south east. However, it has not been as successful as was hoped, having a population of only 1.5 million and attracting few industries. It has certainly not reduced the primacy of Rio and Sao Paulo (both mega-cities).

New capital cities are common in Africa, for example the more centrally located Abuja replacing Lagos as the new capital of Nigeria.

New towns have been constructed to house overspill population and divert growth from the main cities. If new towns are built too close to the main cities they become satellite towns (because people are within commuting distance) or are engulfed by the main city. New towns can act as growth poles. In Singapore the Housing Development Board has been very successful in re-housing people in purpose-built new towns, each with self-contained light industries. In Egypt a number of new towns such as Sadat City have been constructed to relieve pressure on Cairo. People have helped to construct their homes, which they will eventually own.

Rural development is probably the most effective way of easing pressure on the main cities. Rural to urban migration has had a profound impact on rural areas, leaving ageing and unskilled communities behind – sustaining their livelihoods is difficult. The way to slow down rural to urban migration is, of course, to develop the rural areas to make them more attractive. Rural areas need:

There has been little emphasis on rural development.

- appropriate education (agricultural advice to increase productivity and skills to develop small-scale industries) – the Green Revolution attempted to do this but only the wealthy farmers have really benefited because of the investment needed;
- improved healthcare and family planning aimed at women to decrease birth rates and infant mortality;
- improved communications and entertainment;
- land reform to ensure that farms are less fragmented and that single land

owners do not dominate; many land reform schemes have only been partially successful because of lack of government backing;

- the use of appropriate technologies using local skills and materials so villagers can manage their own development;
- government investment in small-scale irrigation and electricity-generating schemes – large-scale schemes such as the Aswan Dam in Egypt and the Three Gorges Dam in China may have benefits but at what cost?

Compare the success of small and large scale projects.

Developing cities will continue to grow in size unless rural areas are developed. In Brazil's largest cities the more wealthy inhabitants are starting to decentralise and move to smaller urban areas in response to overcrowding and pollution. Current growth rates are not sustainable, yet the flow of rural to urban migrants in many countries has only just begun.

Settlements and Urban Areas in LEDCs

45 minutes

Use your knowledge

1 Why do cities tend to have lower birth rates than the surrounding rural areas?

2 Describe and explain the distribution of the world's largest cities. Use the map of the world's largest cities to help you.

3 Describe in detail the environmental problems caused by rapid growth in developing cities. Refer to more than one example that you have studied.

4 Explain why rural development is as important as urban development.

Hint

Consider why people have children.

This is a spatial description question.

Make sure you use case studies

Rural development is the key to reducing rural to urban migration.

Economic Activity

10 minutes

Test your knowledge

1 What are tertiary industries?

2 What does the term 'occupational structure' refer to?

3 What are 'footloose' industries?

4 What is 'de-industrialisation'?

5 What are TNCs or MNCs?

Answers

1 Tertiary industries are involved in providing services. They may be consumer or producer services. 2 Occupational structure is the percentage of a country's work force employed in the different sectors of industry (primary, secondary or tertiary). 3 Footloose industries are not tied to raw material and can in theory locate anywhere. They include hi-tech industries. 4 De-industrialisation is the relative or absolute decline in the proportion of the workforce involved in manufacturing industry. 5 Transnational corporations or multinational corporations who have operations in more than one country.

✔ **If you got them all right, skip to page 81**

Economic Activity

Improve your knowledge

1 Types of industry

Primary industries are concerned with the extraction of natural resources (includes mining and quarrying, agriculture, fishing, forestry and hunting).

Secondary industries are concerned with the processing of primary raw materials and the resultant manufacture of products.

Tertiary industries are service industries, whereby tangible products are not made. However, many service industries assist manufacturing industries (producer services).

Quaternary industries are an extension of the service sector. Activities include training and research and development – they are linked to hi-tech industries and financial services.

These are known as sectors of the economy.

2 Employment structure

Like occupational structure, employment structure compares the percentages of workers employed in different sectors of the economy. The percentages employed in the different sectors gives an indication of economic development.

Employment structures in selected MEDCs and LEDCs

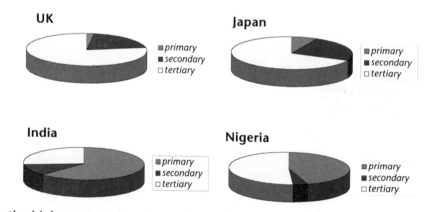

UK — primary, secondary, tertiary

Japan — primary, secondary, tertiary

India — primary, secondary, tertiary

Nigeria — primary, secondary, tertiary

Notice the higher proportions in services and manufacturing in MEDCs and higher percentage in primary (farming) in LEDCs. The service sector in LEDCs can be attributed in part to the informal sector.

3 Factors affecting the location of present-day manufacturing industry

- Raw materials are no longer as important because they can be transported and imported more easily; industries are footloose if they do not rely on raw materials (hi-tech industries along the M4 corridor).
- Power supplies are less important because electricity is the main energy source.
- Transport is an important factor – proximity to motorways, airports and deep water ports are important.
- Markets are very important in terms of demand for the product and the bulk or perishable nature of the product.
- Suburban locations tend to be chosen because land is cheaper, there is more room for expansion and sites are chosen in business parks next to motorway junctions. Industries agglomerate to save and share costs.
- Labour costs are important, as is the availability of skilled labour for some types of manufacturing.
- Government policies are a major factor. Enterprise Zones with lower taxes encourage location, whilst strict environmental laws may deter location.
- TNCs dominate world trade, and decisions made about location by these companies have a major impact.

South Wales produces iron and steel but imports all the iron ore and most of the coal.

TNCs have moved assembly to LEDCs where labour costs are lower.

4 Industrial development

W Rostow put forward a simplistic model of economic development based on the growth of Western economies during the Industrial Revolution. Most LEDCs have failed to follow the same path to economic growth.

Rostow's model of economic development

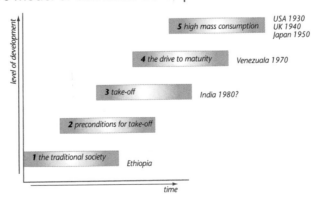

Economic Activity

The newly industrialising countries (NICs) have achieved industrial growth in manufacturing by import substitution (producing goods internally rather than importing them) and through government in investment in export-orientated growth. The creation of export processing zones and free trade zones has helped to promote this growth

The first wave of NICs started in the 1960s, namely Singapore, South Korea, Taiwan and Hong Kong. A second wave of NICs includes Malaysia, Mexico and Brazil.

New methods of production have accompanied industrial change, including the Japanese 'just in time' method.

Industry in LEDCs is mainly concentrated in urban areas, particularly the primate city.

4 Changing employment structures

MEDCs have suffered from a loss of employment in manufacturing in the face of overseas competition. This does not mean that their manufacturing outputs have fallen dramatically; rather that a higher proportion of people are working in service industries. This process is known as de-industrialisation. The sector model identifies how different sectors change in size and relative importance over time.

The old industrial regions of the UK have suffered most from de-industrialisation whilst the south east has grown most in terms of hi-tech industry and services. The north-south divide is often quoted.

The sector model

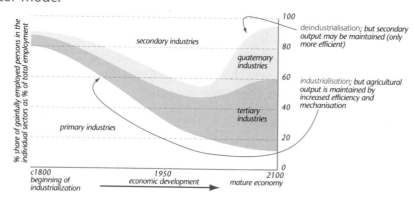

5 Globalisation of industry

The globalisation of world industry has resulted from the integration of world trade, improved communications, increased competition, rising demand, and foreign investment by TNCs.

Economic Activity

30 minutes

Use your knowledge

Hint

1 What factors would influence the location of a hi-tech industry?

2 What could be the economic, social and environmental impacts of the closure of manufacturing industry in an area?

3 What might be the advantages and disadvantages of a TNC setting up a branch plant in an LEDC?

Remember this is a modern industry.

Impacts may be positive or negative.

Try to give a balanced answer.

60 minutes

 Study the diagram below that shows the long profile of a stretch of upland river.

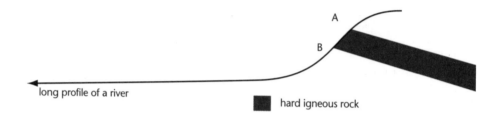

A

B

long profile of a river

hard igneous rock

a) Describe and explain the formation of channel features which are likely to develop between points A and B. (5 marks)

b) Which parts of the river's course are likely to be dominated by erosional features and why? (5 marks)

 a) Define the term 'watershed'. (2 marks)

b) How can the position of the watershed of a drainage basin be identified? (2 marks)

c) Describe how the discharge of a river would be affected by it flowing through an urban area. (4 marks)

3 Study the table below that compares the catchment areas of two rivers:

catchment area details	river X	river Y
rock type	mostly impermeable (granite)	mostly permeable (limestone)
maximum altitude (metres above sea level)	600	189
mean annual precipitation (mm)	1125	660
mean annual run off (mm)	820	168

a) Describe the different relationships between mean annual run off and mean annual precipitation. (2 marks)

b) Why is mean run off lower in river Y than X? (4 marks)

4 a) Explain where and why intense convection occurs. (2 marks)

b) How might an urban area that is susceptible to flooding reduce the likelihood of damage from flooding? (6 marks)

5 Study the following table:

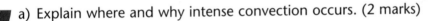

Habitat description	Reed swamp	Marsh or fen	Open wooded fen	Closed wooded fen	Woodland
Number of plant species	5	11	15	28	14

a) Why are there fewer species in the woodland stage than in the previous stages? (2 marks)

b) Explain how human and physical factors may cause a plant succession to be arrested. (8 marks)

Study the population pyramids below:

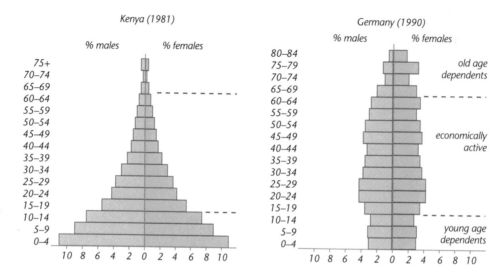

a) Define the terms 'economically active population and 'dependent population'. (2 marks)

b) Why might the age groups of the groups mentioned in (a) differ between countries? (4 marks)

c) Why does the population structure of Kenya cause rapid population growth rates? (4 marks)

d) What are the implications of an ageing population structure as shown by Germany's pyramid? (4 marks)

e) How might people attempt to resolve the problems caused by an imbalance between population and resources? How successful have they been? (10 marks)

Exam Practice Questions

 7 Study the diagram below that shows a model of a CBD in an MEDC city.

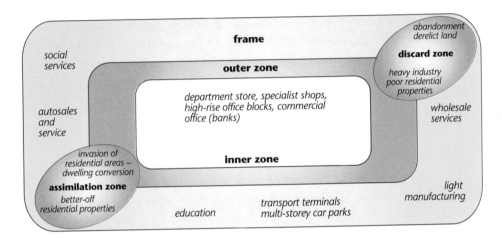

a) Define the term 'CBD'. (2 marks)

b) Explain the distribution of functions within the CBD. (4 marks)

c) How might future planning policies utilise land in the discard zone shown on the diagram? (4 marks)

8 a) Define the term 'urbanisation'. (2 marks)

b) Discuss the causes of rapid urban growth in many developing cities. (4 marks)

c) Discuss the success of government policies to address the problems caused by rapid urban growth in LEDC cities. (8 marks)

Answers on page 94

River Basins and Hydrology

1 Vegetation increases infiltration capacity by slowing infiltration rate. Primary and secondary interception, throughfall and stemflow (explain these) lead to less run-off. Protects against raindrop impact and soil compaction which lower soil's capacity. Loss by evapotranspiration reduces chance of the river reaching bankfull stage.

2 a) 12 hours

b) 10 cumecs

c) Make comparative statements (e.g. A has a more shallow rising limb than B) about the differences in peak discharge, lag time, baseflow and steepness of both limbs.

d) The basin is probably elongated with few streams. It is likely to be densely forested with gentle slopes and permeable soil. Expand on these and say how they delay the water getting into the river.

e) Much shorter lag time and higher peak discharge. Explain how water gets into channels quickly due to impermeable surfaces. Explain how lack of vegetation affects interception, infiltration and so on.

3 Explain it is a natural process worsened by man. Refer to deforestation and urbanisation. The question invites you to deal with degree (increasing the scale of a natural flood) and frequency (lower recurrence intervals) separately. Describe ways of reducing flooding such as dam construction and refer to floods that you have studied such as Mississippi (USA) 1993 or Yangtse (China) 1998.

Coastal Environments

1 Ideally refer to specific stretches of coastline which show signs of both, such as East Yorkshire or Northumbria. Mention weathering assists erosion. Emphasise link between eroded material and depositional features (shingle at Spurn Point from Holderness). Try to include a sketch map of chosen coastline and one of the individual feature.

2 Only very brief reference to wave action required. 'Other factors' are geology / rock structure and weathering. Refer to dips in strata, well-jointed open structures, and rock hardness. Describe how these affect cliff form and shape (hard rock gives steep cliffs). Explain various weathering processes acting on upper cliff and how these can accelerate wave action and influence cliff form.

3 a) X = compound spit, Y = salt marsh

b) Low-lying mud, covered at high tide, halophytic vegetation, often bisected by creeks at low tide.

c) Halophytic area is inundated twice a day. A succession of vegetation types – a halosere. Very salt-tolerant plants (e.g. samphire) in lowest areas, less tolerant species higher up (e.g. sea asters).

d) Salt marsh is a low-energy environment. Mud builds up around plant roots and more species become established. Plants die and add organic matter. Soil depth increases allowing other species to become established. Describe in detail, naming species. You could mention how such areas are often artificially planted for stabilising the sediments or drained for agriculture.

4 Explain eustatic changes (due to glaciation, deglaciation and changing oceanic capacity due to sedimentation), and isostatic changes (due to pressure-release after deglaciation and tectonic processes). Mention possible future changes due to global warming. Refer to specific examples of landforms and say how man is influenced by them or has adapted to them: rias/fjords (make ideal ports); islands (settlement/transport problems); indented coasts (need for ferries or bridges); tourist potential of landforms; raised beaches (ideal farming land and natural routeways).

Ecosystems

1 Your diagram should have the three different sized circles: a large biomass circle and small soil and litter circles. There are large flows of nutrients from soil to biomass, medium flows from litter to soil and leaching from the soil, and small flows from biomass to litter.

2 Break in nutrient cycle as major store (biomass) is removed. Litter store will disappear, soil store will be leached away. Less nutrient input from reduced precipitation due to reduced evapotranspiration. Consequent loss of soil fertility and reduced species diversity. Deforestation reduces interception, encourages rapid run-off, soil erosion and flooding.

3 a) Litter is fallen plant and animal material, mostly leaves, which forms a layer on top of the soil. It is rich in nutrients and is incorporated into the soil once decomposed. Biomass is the total mass of all living plant and animal matter, also a store of nutrients.

b) Litter is decomposed into humus by bacteria, fungi and some insects. It is incorporated into the soil by earthworms and burrowing insects.

c) Forest has small soil store due to low incorporation by insects due to low temperatures. Prairie has high soil store, as extensive roots are underground. Expand on these points.

Tectonics and Surface Processes

1 Constructive margins (Mid-Atlantic Ridge): two oceanic plates diverge; submarine volcanoes along its length may grow above sea level. 1963: Surtsey, near Iceland. Destructive margins (Nazca and S. American plates): oceanic crust is subducted under continental crust. Friction produces heat – melts the crust. Rising magma reaches surface to form volcanoes such as Cotopaxi.

2 Immediate events – falling lava or collapsing buildings. Secondary effects: first few weeks – fire, disease and water-related problems. Long-term effects: how the communities will be affected over years. Minimising effects: ways of improving prediction, warning and evacuation; diverting lava flows or adapting buildings.

3 High mean annual temperature increases chemical reaction time, so chemical weathering is effective (describe each) and rock is weathered to a greater depth. Plentiful precipitation also increases chemical weathering. Lush vegetation increases biological weathering by organic acids. Chemically stable minerals may remain but some (limestone) are very deeply weathered.

4 a) Write two or three lines of careful definition – see 'Improve your knowledge'.

b) Mechanical weathering, especially exfoliation. Low chemical weathering due to low precipitation.

c) Diurnal temperature fluctuations around 0°C produce the greatest number of expansions; polar regions only have one freeze-up per year in autumn and therefore few expansions. Water is frozen so cannot trickle deeper into cracks.

d) Define carefully – see 'Improve your knowledge'.

e) Flatten angle of slope by terracing or benching; improve drainage so that mass does not increase as much; fix steel mesh or steel pilings to stabilise vertical rock; re-route heavy traffic.

Glacial Environments

1 a) Good, clear detailed definition – see 'Improve your knowledge'.

b) Decrease in temperature: More precipitation falls as snow so more accumulation. Ablation is lower so net increase in ice mass – a positive budget. Decrease in precipitation: Reduced accumulation at all times of year but especially in winter. Summer ablation rates are unaltered so there is a negative budget and the ice retreats.

2 a) See 'Improve your knowledge'. Use a clear, labelled diagram.

b) Freeze-thaw on backwall causes rockfalls and build up of scree in a Talus cone. A corrie tarn or small lake may occupy the basin. Moraine dumped on sides and lip of corrie becomes rounded. Moraine on lip may be cut through by a tarn stream.

c) There will be depositional features at the ice margin, mostly terminal moraine. There may be an outwash plain extending away with drumlins and eskers. There will be a transition from unstratified (ice-dumped) to stratified (fluvio-glacial) deposits.

Meteorology and Climate

1 a) Condensation level = broken line at 1 km altitude, DALR = portion of solid line below condensation level. SALR = portion of solid line above condensation level.

b) On hot days, warm rising air stays warmer than surrounding air. If dew point is reached, uplift is accelerated to form storm clouds with thunder, lightning and heavy precipitation. Uplift and instability can occur in the mid-latitude depressions which affect NW Europe and in tropical cyclones (very deep, powerful depressions). Describe the conditions associated with each.

2 a) i) Transfer of energy by movement of 'parcels' of air. ii) Energy is used when liquid water is changed to vapour, creating a localised loss of energy.

b) i) Reduced significantly. ii) Reduced as less solar radiation is received. iii) Less will be outgoing as surface temperature falls but more will be reflected back by clouds. iv) Less is absorbed since less is available.

c) Car park has less stored heat and is likely to be more exposed to the wind causing greater heat loss. Car park is a better radiator of heat so greater long wave radiation loss.

Population

1 Discuss physical and human factors that explain population distribution. Physical factors include climate, vegetation, relief, water supply and resources. Human factors include government policy, industry, farming, and communications. Write about a named region and show you have a sense of place. Mention how the factors may have changed over time and space. An annotated map may save long descriptions. Try to mention anomalies that do not fit the pattern.

2 Population characteristics refer to age structure and sex as well as acquired characteristics such as education and religion. For your case study country, know birth rates, death rates, infant mortality, population totals and percentage growth rates. Explain how and why the country has moved through the stages of the demographic transition.

3 Overpopulation means too many people. According to Malthus this would be when population outstripped food supply and resources. In other words the carrying capacity is exceeded. Examples are parts of China and Bangladesh. Give a definition, theory and examples.

4 Use case studies to illustrate the indicators that suggest overpopulation (famine, environmental degradation, contaminated water supplies, unemployment, poor housing, low standards of living, poverty, etc.). Remember that uneven distribution of wealth may intensify the problem. Also technology (Boserup's theory) may help to overcome the problems (Green Revolution for instance) and that carrying capacities can be increased over time. Some estimates suggest that the world could support 35 billion people, but at what standard of living?

5 The physical environment refers to the environment and how it may become degraded. One example could be the Sahel in Africa where increased population has caused over-cultivation, overgrazing, deforestation, soil erosion and desertification. Notice the links here with ecosystems and hydrology.

Migration

1 a) Barriers could be immigration laws or an individual's lack of capital or skills

b) Governments put barriers in place to prevent large influxes of population putting on a strain on resources, the economy and the environment. Also countries do not want to lose skilled people. People may not be able to return after making the decision to leave if they have left on grounds of persecution.

2 Remember to split your answer by classification of migration giving an example of each; forced or free and temporary or permanent. Causal push and pull factors are physical (hazards, climate,); economic (jobs, standard of living); social (education, family) and political (persecution, war). Try to conclude by highlighting the most significant factors and mentioning the factors that will affect how far an individual can migrate.

1 Refugees are people who are forced to leave their home country due to political or religious persecution and who are in fear of returning. Refugees may cause environmental impacts to the home and host countries. Environmental impacts refer to land and resource degradation, e.g. in refugee camps. Social impacts include effects on age structure and ethnic tensions.

Settlements and Urban Areas in MEDCs

1 On your sketch map include factors related to the site itself. Explain how they have influenced the choice of location and influenced future growth. Some factors may no longer be relevant; this adds a time dimension.

2 Central places exert an influence whatever their size. Use your areas to demonstrate why different sizes of settlements have different catchment areas. Use theories like Reilly's to discuss breaking distances and mention physical barriers or human factors that may increase or decrease the spheres of influence of some settlements. A map is always useful.

3 Do not attempt to fit a town or city to an urban model. Rather look for characteristics that are applicable and explain how useful, or not, the models are in explaining the land use in a settlement.

4 All the issues may be positive or negative. Socially, there may be more community developments and better facilities. However, gentrification may occur or housing may not be suitable (flats for families). Environmentally, polluted areas may have been cleaned (dockland water) but new schemes may not provide sufficient open space. Although jobs may have been brought to an area they may not suit local skills or they may be transferred jobs from elsewhere.

5 For: easy access for customers and employees, relieves congestion of crowded CBDs, space for parking, pleasant environment, construction provides jobs as do new businesses, may prompt more developments nearby.

Against: encroaches onto greenfield sites and perhaps areas of scenic or natural interest, encourages car usage instead of public transport, competes with the CBD and causes closures of shops, difficult access for those without cars, jobs are often temporary or transferred

Rural Areas in MEDCs

1 Functions may have become diminished due to commuters using urban facilities, or may have become more urban if the village has become suburbanised. This will be true if there are families living in the village who demand schooling, shops, leisure facilities etc.

2 Counter-urbanisation is a result of push and pull factors. Write about the negative issues in cities and the positive ones in smaller settlements. Think in terms of environmental, social and economic factors. Mention the age groups who are most likely to move.

3 a) Wealthy people may have bought second homes, holiday homes, or retired to the areas. People may have renovated the properties and sold them on at a higher price. The increased popularity has increased the prices forcing local people out. Examples are villages in the Lake District (Braithwaite) and Peak District national parks (Castleton).

b) Social impacts may mean that new generations of people are unable to afford houses in their own villages and may be forced to move elsewhere. This may have an impact on the economy because the new residents tend not to work in the village. There may be a loss of population to carry on trades or do rural jobs like farming. The village may lose its character and become urban in function or may have few functions because they are no longer supported. Try to mention case studies.

Settlements and Urban Areas in LEDCs

1 Birth rates tend to be lower because there is less need to have children to work (in farming) and infant mortality is lower so more children will survive. The status of women may improve and they may make decisions to have fewer children.

2 Largest cities used to be concentrated in Europe and North America but there has now been a shift to the 'poor South' (continents of Asia, South America and Africa). Most of the largest cities are now located in LEDCs, many in Asia. There has been a shift from over 40°N to the area around the tropics.

3 Environmental problems include pollution of air and water, land degradation, refuse collection, etc. you need to refer to one or more cities (pollution in Mexico City, land slides in Rio de Janeiro). Explain how they have come about and why they are becoming worse.

4 Rural development will stop people moving to the cities. Remember that cities are often a drain on the country's resources and perceptions of city life are often false. Try to justify why rural development is more sustainable and why government money may be more effectively spent in the rural areas. Try to refer to case studies to support your answer.

Economic Activity

1 Hi-tech industries are not tied to raw materials, but they need excellent transport links and proximity to financial services. Research and development will be concentrated near cities with universities and highly skilled workforce such as the Cambridge Science Park. Production will be near markets, but also in areas where costs can be minimised, possibly in LEDCs (in Export Processing Zones).

2 Social and economic impacts will be negative unless there is government investment and relocation of new industries to the area. Environmental pollution may be reduced or cleaned up to attract new industries.

3 TNCs bring new investment and hence jobs; this will improve skills and technology. It may bring products to the market and expansion of service jobs. However, jobs may be low-paid and skilled workers are imported. Profits are repatriated to the home country and environmental legislation maybe lax. The company could withdraw at any time.

1 a) Features likely to be present: waterfalls, rapids, gorges and potholes. If the band of hard rock is dipping upstream, the softer rock underneath is undercut by the river at A, forming a waterfall. As erosion and undercutting continues, the hard rock collapses and the river retreats forming a deep gorge. Potholes are formed by abrasion, and these join up to lower the river's bed. If the hard rock dips downstream, rapids will form where the river erodes the bed slowly.

b) Erosional features are more likely to be found in upland areas near the river's source. Such features are caused by vertical erosion and sometimes by lateral erosion in the case of meanders. Explain how steep gradients in upland areas increase the river's energy and so the rate of erosion. Describe erosion types (corrasion etc). List and briefly describe some erosional features (waterfalls and truncated spurs).

2 a) The boundary line which separates the headstreams which are tributary to different drainage basins. As rivers can 'capture' other rivers, the watershed is not permanent. It often follows the mountain ridges which separate two valleys.

b) This can be found by locating the points at which the flow directions of rivers change. On a relief map, the contours can show the ridge tops which separate drainage basins.

c) Discharge may be partially reduced by the use of river water by industry in the town. However, the smooth, efficient drainage system would increase the speed water reaches the river, so increasing discharge. Impermeable urban surfaces and lack of interception by vegetation should also be mentioned.

3 a) There is a greater difference between average precipitation and average run off in catchment area Y. Quote figures from the table. Calculate the difference between the two values for both catchment areas.

b) Rock permeability: limestone allows percolation through joints and bedding planes; Y has lower precipitation so more chance of the ground being unsaturated, so infiltration would be higher.

4 a) Occurs in continental interiors in summer – give examples (central USA, central Russia) – and in tropical regions all year. Intense convection also occurs in areas subjected to severe weather such as storms and tropical storms called cyclones, typhoons or hurricanes.

b) Banks can be made higher; channels can be made more efficient by straightening them (reducing sinuosity); smooth concrete channels are often built. Channels must be kept clear of urban debris such as old bikes. Refer to examples in your town. How has the river's course been altered? Note that you cannot write about flood abatement strategies such as afforestation in this question.

5 a) Woodland has less dense surface vegetation due to shade. Greatest number of species at 'closed wooded fen' stage due to open foliage and abundance of water-edge species.

b) Natural arresting factors which could halt the development of a plant succession (prisere) include volcanic activity when ground is covered by ash or lava; mass movements such as flows, slumps and slides and erosion and deposition by rivers, glaciers or the sea. Explain in each case how the land is covered to form an untenanted site. Human factors include overcropping, overgrazing and clearance, sometimes using fire, for settlement, extractive industries or agriculture.

6 a) Economically active are the adult working population aged 15–64. Dependent population includes children (0–14) and elderly (65+) who rely on the active population.

b) Age groups differ because in LEDCs children may work from as young as 5; in both MEDCs and LEDCs people work beyond retirement age and contribute to the economy. Many jobs may be informal in LEDCs.

c) Kenya has a young age structure with many children; they have the potential to reproduce. As infant mortality and life expectancy increase more people will survive in each age group therefore increasing population. Fertility rates are likely to be high and even if they fall there are many young people to have children.

d) Implications of an old age structure include pressure on services (health care, sheltered accommodation) and too few economically active people to support them. This has further implications for pensions and the amount that the state can provide. It may mean that the retirement age must be raised.

e) Policies may be related to resources: improving agriculture (new crops, Green Revolution in India), industrialisation (NICs), encouraging or discouraging migration (Brazil), finding or creating more living space (Japan); population – limit growth via birth control (voluntary or government policy, e.g. China), education and increasing the status of women. High level answers will evaluate the policies and discuss positive and negative impacts and refer to specific case study material.

7 a) CBD is the central business district. It is the area containing the high order retail, administrative and commercial functions of an urban area. It is in theory the most accessible area, therefore land values and building density are high.

b) Bid rent values will decrease with distance from the Peak Land Value Intersection (PLVI), therefore the inner zone will contain the functions which can bid the highest rents. Functions requiring more space will locate near the edge of the CD into the twilight zone. The highest bidders may change over time.

c) Brownfield sites are left when a building has been abandoned or demolished. It is current UK government policy to redevelop these areas in preference to greenfield sites on the edge of urban areas. Developments may include high-class housing, offices, retail parks etc. Give case studies if you can.

8 a) Urbanisation is an increase in the proportion of people living in towns and cities. It is therefore a relative as well as an absolute increase.

b) Rapid urban growth is caused by rural push and urban pull factors, which you could describe with reference to a named case study. In addition natural increase is high because migrants are likely to have larger families because rural traditions are still strong.

c) You need to evaluate the success of self-help, site and service schemes and tower blocks by using case study material. Discuss economic, social and environmental benefits and problems. Make sure you get a perspective of scale and decide whether government policies or simply the initiative of migrants themselves have been more effective. Have some policies of bulldozing shanty towns actually been detrimental? Question whether any of these solutions actually tackles the cause of the problem and whether rural development may be more successful.